Keeping All the Pieces

Perspectives on Natural History

and the Environment

Keeping

Smithsonian Institution Press • Washington and London

All the

PIECES

Whit Gibbons

Editor: Catherine F. McKenzie
Production editor: Jack Kirshbaum
Designer: Linda McKnight

Library of Congress Cataloging-in-Publication Data

Gibbons, Whit, 1939–
     Keeping all the pieces : perspectives on natural history and the
environment / Whit Gibbons.
        p. cm.
     Includes index.
     ISBN 1-56098-224-1 (alk. paper)
     1. Ecology. 2. Natural history. 3. Human ecology. I. Title.
QH541.13.G53   1993
574.5—dc20   93-2037

British Library Cataloguing in Publication Data is available.

Manufactured in the United States of America
97 96 95 94 93   5 4 3 2 1

∞ The paper used in this publication meets the minimum requirements
of the American National Standard for Permanence of Paper for
Printed Library Materials Z39.48-1984.

To Anne R. Gibbons
with lasting appreciation

To keep every cog and wheel is the
first precaution of intelligent tinkering.
                                    —Aldo Leopold

Contents

Foreword

The diversity of life on this earth of ours is incredible, and we must not take it for granted. We do assume, on pretty good ecological evidence, that this abundance is itself important in maintaining the many processes upon which humans, and all other creatures, ultimately depend. It is this life, and its ubiquity, that distinguishes Earth from nearby planets, where humans cannot survive without some kind of artificial life-support system. The full range of life on Earth *is* our own life-support system, and we must protect it.

We know that there is considerable redundancy and resilience in the fabric of nature on Earth. In past ages, comet impacts, glaciers, volcanic eruptions, and other geologic events have wrought terrible destruction on species and ecosystems. Each such event has been followed by recovery of biodiversity, the evolution and diversification of new forms on a time scale of tens of millions of years. We worry now that the increasing abundance and destructiveness of humans threaten to do more permanent damage to Earth's networks of life than any past geological event.

Solutions may still be available through the application of basic ecological principles. Ecology is an integrative science. It attempts to link knowledge of the physical world with the study of life, with the goal of understanding the full interrelationships of ecosystems and the whole biosphere. Today, in the context of conservation, ecologists are often asked to apply their knowledge to assessing an ecosystem's "health" and to prescribe procedures for "restoration." Just as a thorough grounding in anatomy and physiology is essential for healing the human body, a fundamental understanding of ecology is essential for interpreting the patterns and processes of nature, and moving toward an appropriate cure.

One unfortunate but common obstacle to understanding and addressing problems is their complexity and the difficulty that scientists have in communicating their concerns clearly to a broad public. Whit Gibbons bridges this gap. He is a trained research ecologist of international renown, but he is also a very gifted storyteller. He can view the world with a practiced ecological eye, cognizant of subtle effects and relationships, but in his descriptions never loses the enthusiasm for life that we all have felt. His writing arouses in us again that fundamental enjoyment of nature. He tells us of panthers and primroses, snakes and turtles, insects and birds, and we emerge with a greater understanding, and greater respect, for our natural world. Here we have a natural history at its best, with a human message.

Eugene P. Odum
Director Emeritus, Institute of Ecology
University of Georgia

Preface

Think of the world's ecosystems as brick buildings in which each brick represents a different species. Plants and some animals form the base of a building. The position and stability of each depend on those below or, in the case of keystones, those above. Larger consumers are toward the top. When one brick is removed, the building does not fall, but the structure becomes weaker and less stable. When several are removed, the remaining bricks begin to shift position. Certain key bricks will have to be removed before the entire building falls; but the removal of one critical brick may cause a wall or an arch to collapse. As other bricks are removed, we soon have a pile of rubble.

Our environmental problem today is that we are removing bricks without knowing where they fit into the structure of the building, without regard to their interdependence. Bricks are falling all around us. Ecological rubble piles are being created everywhere, in the United States and throughout the world, and at a disturbing rate. Somewhere near the top of a building is a brick that represents the human species. Our brick may be one of the last to fall, but we won't have a very pretty or useful building to look at while we wait.

Just as no single brick holds a building together, so no single organism is the critical one upon which our existence hinges. Each species in our environment, no matter how small or seemingly insignificant, is part of the overall supporting structure that carries us through time. And whether we know anything about the species or not, its place in the world affects us. Maybe indirectly, maybe only occasionally. Possibly through unidentified interaction with another species. Its effect may not even be perceived by this generation of humans. But every animal and plant species has its unique significance to its own world—and to ours.

This book is about some of the many species that take part in holding up our world, ecologically. Some are extinct, some are endangered, others are doing well. We are depending on the extant ones, and each time one is lost, an adjustment, however slight, must be made. We have a long time left as an extant species; to live well and to live up to our responsibilities as the most intelligent of the species, we must respect and maintain the other species, upon which we depend.

Of my three goals in this book, the primary one is to foster a deeper appreciation of our natural world—an appreciation that is expressed superficially every day when people tend their houseplants and pets, maintain shrubs and home gardens, go canoeing and camping, take walks through the woods.

My second goal is to affirm the intricate and delicate nature of the relationships and interactions among plants, animals, and their environments by offering a nontechnical yet detailed perspective of natural history. Experts in a given field may perceive some of the presentations as elementary. But the experts are the sources for this book, not the audience.

My third goal is to identify human attitudes and actions that can lead us in either of two ways: toward environmental degradation on a planetary scale or toward a stronger communion between humans and nature. Why are so many people ignorant of and resistant to the importance of environmental preservation? Who are the culprits of environmental abuse, and why do they participate in it? Once culprits are clearly identified, what can we do to redress the problems? In our quest to make the world better suited to our own needs, we often interfere with the ability of other species to meet their needs. An accounting of environmental disruption, both past and present, including a look at modern extinctions, gives insight into mankind's effect on the environment.

I believe we should appreciate the singularity of every species on earth as well as the collective value of these species in forming natural systems. Increasing our awareness and familiarity with the intricate world of nature will ensure that we place higher values on our natural heritage and defend it against forces that would destroy it. As Thomas Berry says in *Dream of the Earth,* "A degraded habitat will produce degraded humans."

Some might argue that humans can devise ways to survive without the natural world, that we need not depend on it for our fundamental existence. Perhaps so, but is this the world we *want* to live in? People will find no joy in a world without wildlife.

Acknowledgments

I owe thanks to many people for help in compiling natural history accounts of plants and animals. Most of these were originally presented in my newspaper columns in the *Aiken Standard* of Aiken, South Carolina, and the *Tuscaloosa News,* Tuscaloosa, Alabama. I thank the staffs of these papers for providing an outlet for the original publications. I particularly thank Ken Stickney of the *Tuscaloosa News* and Anne Gibbons for assuring a more widespread distribution of the columns.

This book could not have been completed without the dedicated assistance of several individuals who helped track down reference material or performed clerical tasks. I especially appreciate the unfailing assistance of Sarah Collie, Marie Fulmer, and Judy Greene of the Savannah River Ecology Laboratory.

I also thank those who read and offered comments on particular sections of the manuscript. Their suggestions were much appreciated as they identified many commissions and omissions that needed attention. I especially thank Barbara Dietsch, Carol Gibbons, Janie Gibbons, Bruce Grant, Judy Greene, Joe Pechmann, and David Scott for their help.

I also appreciate the time invested by others, including numerous professional ecologists, who read sections relating to their areas of expertise and corrected my errors in interpretation of the findings. In particular, I thank the following: Vikram Akula (Worldwatch Institute), Richard D. Alexander (University of Michigan),

ACKNOWLEDGMENTS

Randy Babb (Arizona Game and Fish Department), Brian M. Barnes (University of Alaska, Fairbanks), Stephen H. Bennett (South Carolina Wildlife and Marine Resources Department), I. Lehr Brisbin (Savannah River Ecology Laboratory), Vinny Burke (University of Georgia), Heyward Clamp (Edisto Island, South Carolina), Justin Congdon (Savannah River Ecology Laboratory), David Conover (State University of New York at Stony Brook), Martha L. Crump (University of Florida), Joseph E. Daniel (*Buzzworm Magazine*), Margaret Davis (*Outside Magazine*), Eric Dinerstein (World Wildlife Fund), C. Kenneth Dodd, Jr. (U.S. Fish and Wildlife Service), Phillip D. Doerr (North Carolina State University), Earthwatch Expeditions, Inc. (Watertown, Massachusetts), Earthworks Group (Berkeley, California), James Ehleringer (University of Utah), Thomas Eisner (Cornell University), George Folkerts (Auburn University), Nigel R. Franks (University of Bath, United Kingdom), Alan R. French (State University of New York at Binghamton), Graduation Pledge Alliance at Humboldt State University, Erick Greene (University of Montana), Paul Haemig (University of Umea , Sweden), Curt L. Harbsmeier (Central Florida Herpetological Society), Rachel M. Haymon (University of California, Santa Barbara), Edward J. Heske (Illinois Natural History Survey), R. Howard Hunt (Zoo Atlanta), John Irwin (Savannah River Forest Station), John Iverson (Earlham College), Ted Joanen (Louisiana Department of Wildlife and Fisheries), Dennis Jordan (U.S. Fish and Wildlife Service), Jay Keller (Zero Population Growth), Robert A. Kennamer (Savannah River Ecology Laboratory), Erich Klinghammer (Purdue University), James Kushlan (University of Mississippi), Trip Lamb (East Carolina University), James C. Lewis (U.S. Fish and Wildlife Service), Jim Lloyd (Department of Entomology, University of Florida), Jeff Lovich (U.S. Bureau of Land Management), Tracy K. Lynch (Augusta College), William D. McCort (Savannah River Ecology Laboratory), Kenneth McLeod (Savannah River Ecology Laboratory), James B. McClintock (University of Alabama at Birmingham), Bastiaan J. D. Meeuse (Botany Department, University of Washington), Gary Meffe (Savannah River Ecology Laboratory), Bill Melson (Museum of Natural History, Smithsonian Institution), Anne Meylan (Florida Marine Research Institute), Mark Mills (University of Georgia), Tony Mills (Savannah River Ecology Laboratory), Peter B. Moyle (University of California, Davis), Park S. Nobel (University of California, Los Angeles), Ian Parer (CSIRO Division of Wildlife and Ecology, Australia), George B. Rabb (chairman,

xviii

Species Survival Commission of IUCN, the World Conservation Union), James A. Raymond (University of South Alabama), T. D. Redhead (CSIRO Division of Wildlife and Ecology, Australia), Heinz-Ulrich Reyer (Department of Zoology, University of Zurich), Bernard Roitberg (Simon Fraser University), Rich Seigel (Southeastern Louisiana University), Raymond D. Semlitsch (University of Zurich), Dan Sherry (Tennessee Wildlife Resources Agency), South Carolina Department of Health and Environmental Control, James R. Spotila (Drexel University), Peter Stangel (National Fish and Wildlife Foundation), Janet M. Storey (Carleton University), Mark K. Stowe (Gainesville, Florida), Douglas W. S. Sutherland, Stanley A. Temple (Department of Wildlife Ecology, University of Wisconsin), Randy Thornhill (University of New Mexico), Tony Tucker (University of Queensland), Francesca Vietor (Rainforest Action Network), Francisco J. Vilella (U.S. Fish and Wildlife Service), Dick Vogt (Estación de Biologia Tropical, Universidad Nacional Autonoma de Mexico, Los Tuxtlas, Mexico), David B. Wake (University of California, Berkeley), Julie Wallin (University of Georgia, Co-op Fish and Wildlife Research Unit), Peter Weigel (Wake Forest University), Lenore E. West (Midwestern State University), Jeanne Whalen (Greenpeace), World Wildlife Fund, Rebecca Yeomans (University of Georgia), James Young (U.S. Department of Agriculture). I especially thank Nat B. Frazer of Mercer University and Peter Cannell of the Smithsonian Institution Press for their resolute encouragement, guidance, and critical review of material from the initiation of the project to its conclusion. Despite this imposing list of advisors, most saw only short sections, and I did not always take their advice. I thus assume full responsibility for any remaining errors.

Three individuals—Patricia West, Anne R. Gibbons, and Bill Fitts—served in a full editorial capacity, each reading the manuscript in its entirety. I am especially grateful to have had their support. Patricia West spent many hours at the task of editing and made many helpful suggestions in both the early and late stages of the manuscript's preparation. Her friendship, encouragement, and editorial expertise were essential to the completion of the final manuscript. I also owe a lasting debt to Anne Gibbons and Bill Fitts for their editorial vigil throughout the writing process. Patricia, Anne, and Bill also served in an editorial capacity during the preparation of original articles for the newspaper column, so their continued involvement through the process of preparing a book was of special value. I am also greatly appreciative of the suggestions made by Catherine F. McKenzie during the final editorial stages.

Finally, I cannot fully convey my appreciation to my wife, Carol, and our four children, Laura, Jennifer, Susan Lane, and Michael, for their endurance of me during the last-minute deadlines for newspaper articles and the seemingly endless chore of developing them into a book.

Keeping All the Pieces

Prologue: A World without Wildlife

Here's a chilling thought. Consider, from an environmental perspective, what the world could be like a hundred years from now, when the grandchildren of today's kindergarten set are middle-aged. If human population growth and environmental degradation continue at today's rate, here is a grim glimpse of what mankind can look forward to.

The twenty-billion-plus people on earth, four times more than we have now, will inhabit cities that span continents. Can you envision the natural environment in such an urban world? Except for a smattering of wildlife reserves a tiny fraction the size of today's larger national parks, there will be no natural environment. A continuous stream of people who wish to see what the world was once like will visit the few remaining sanctuaries of natural habitat. These token wildlife reserves will be small and disconnected, separated from one another by industrial parks and urban sprawl, vast areas adapted exclusively for human use. On a night flight from Greater New York–Atlanta to Phoenix–Los Angeles, passengers will nowhere see an unlighted countryside. The population of the United States will number over a billion people.

Outdoor recreation will be sharply curtailed. Boating on streams and rivers will be impossible because of industrial wastes deposited in the waterways. Urban development will have swallowed up camping areas. Swimming will be confined to indoor pools. Contaminated waters and harsh ultraviolet rays will prohibit trips to the beach, lake, or outdoor pools. Land area not inhabited by people will be, of necessity, devoted to agribusiness. Cattle ranching will be at such a premium that most people will be forced to consume synthetic meat substitutes or become vegetarians. Only a wealthy few will eat real meat, perhaps in secret.

Most people will satisfy their innate desire for wildlife experience through video. Many of the superinteractive video channels will be devoted exclusively to nature shows. Movie houses will have special nature shows such as "When Big Cats Prowled" or "Wolves in the Wild." Only those children who have paid careful attention in school will be certain whether it was tigers or *Tyrannosaurus rex* that roamed the earth before the dawn of man.

The seas will be dead. The oceans will have no whales, no dolphins, no big animals at all except those able to live on algae that can adapt to pollution. Sea creatures not already exterminated by overharvesting will have to survive in a waste-filled ocean. No tropical rain forests will exist, and people will wonder if habitats with such a diverse array of plants and animals were the product of someone's imagination instead of a past reality.

Those zoos and public aquariums that can survive the economic pressure for housing and agricultural space will be crowded with people who want to see life-forms that used to exist in the wild. The bigger species will be able to survive only in such man-made havens; their natural habitats will be gone. Each year a few more of these large species will become extinct as a consequence of small populations that will succumb rapidly to disease or the inability to replace their numbers through captive breeding. Hunting for game will be a sport of the distant past. People will view most of the larger animals of today's world only through the glass of a museum case. Many of the smaller species of organisms, such as rare tropical plants and insects, will have disappeared without a trace.

Of course, some animals will survive. Rats and roaches do well in urban environments. Fleas and flies will thrive. The few "large" wildlife species that people will be able to see outside of a zoo or museum will include raccoons, opossums, and gray squirrels. All of these can live in urban areas without being on animal welfare programs. Most of them will become pests. Farmers will wage unending war on an ever-evolving community of insect pests that make their living by stealing human food—food that has become ever more precious.

Plants will fare slightly better because the seeds of some species can be readily stored in a viable condition. But forests of huge trees will be only a memory. To say that trees are a renewable resource is fine, if people don't mind waiting two centuries to replace some of them. Imagine, in a public arboretum, a thirty-foot-tall redwood or Douglas fir with a diameter of only a few inches. Beside it is this notice: "Once upon a time trees like this one were so large in diameter that even six people standing side-by-side in front of one of them could not obscure its trunk. Entire forests of trees that big once occupied the northwestern United States—before we cut them all down."

3

Are you tempted to dismiss all that as the whining hyperbole of an environmentalist? Unfortunately it is not hyperbole. It is merely an extrapolation, a projection of the future based on present trends. Unless we alter our environmental course, the scenes I've described are an all-too-possible reality. Human population levels will probably never reach such extremes for reasons beyond our control—famine, disease, war. But we can avoid this grim reality. Some ways to do so are addressed in this book.

PART ONE

The Natural Complexity of Species and Their Relationships

Species' relationships with each other and with their environments are remarkably complex. Without a thorough knowledge of such relationships, we do not know, when we disrupt an ecosystem, whether we are irreparably damaging the interactions within that ecological system. Bizarre traits of unfamiliar and exotic organisms intrigue people, and learning about such marvels can help inspire a regard for our natural environments. But unusual natural history traits, fine-tuned through gradual evolutionary developments, can be found in the most common species. Developing an appreciation of even the most familiar plants and animals can also be an effective way to develop environmental awareness.

We know the intricacies of behavior, ecology, and adaptation for only a micro-percentage of the species on earth. Although more than a million species have been named, that is all we have done—named them; we do not understand them. Many more have not even been named. Of those we know something about, many have complex life history patterns and fragile interactions with their physical, chemical, and biological environments. Presumably, the same complexity prevails for species whose ecology remains unexplored.

Numerous rationales and justifications exist for preserving the endless array of intricate biological relationships and delicate ecological balances. Undescribed or ecologically unexplored species of plants, animals, and microorganisms may harbor yet undiscovered medical secrets. Some species may have potential as new foods or new energy sources. Each gene within a species is a unique storehouse, a treasure chest of biochemical information. Other species may already serve, unnoticed, as biological controls of pest species, as essential biological partners with species we deem important, or as ecologically important detoxifying agents.

As Paul R. Ehrlich of Stanford University and Edward O. Wilson of Harvard University point out, natural ecosystems provide us with certain essential services

6   that we depend on for our existence. Collectively, the living organisms of the world create an atmosphere with the proper mix of gases to sustain human life. The tropical rain forests alone have a global influence on climate. Plants and microorganisms worldwide are critical in the manufacture and maintenance of soils. Reducing the number of species in an ecosystem diminishes its effectiveness in performing these roles. And we seldom know which species are most critical for maintaining the integrity of an ecosystem. But the most compelling justification of all for keeping all the pieces is that each species has its own intrinsic value, its own right to existence.

The Primrose Path
of Ecology

Nature shows and books pique our interest in animal ecology and behavior, but plants can be equally captivating. A few even have behavior patterns worth noting. An awareness of plants' intricate ecology lets us value them beyond their vital roles as food and fuel. Many ecological questions can be answered if we know the right book to examine or the right expert to consult. However, scientists have barely begun to explain the endless repertoires in the lives of plants and animals. The whys and wherefores of many natural phenomena are still unknown, and new mysteries constantly emerge. But lack of knowledge need not preclude our enjoyment of observing nature.

On a trip to visit my sister, accompanied by neither books nor experts, I observed a fascinating environmental event. I did not know if the phenomenon was well known to plant ecologists, but that did not detract from my enjoyment of the spectacle. As we prepared to sit down for the evening meal at my sister's house, a neighbor, Donna, knocked frantically on the door. "Hurry!" she said. "The primrose is blooming." We looked at Donna standing on the porch, then at each other. An unspoken but obvious question hung over the table: Is this blooming primrose worth returning to a cold supper?

My sister's and brother-in-law's looks were edged with some concern for their friend and neighbor—what was she talking about? Didn't primroses bloom often? But Donna's urgency prevailed. The group decision was made, and the three of us followed Donna across the yard to her house. She headed toward a row of shrubbery. In the fading light of dusk we could just make out each other's sidelong, quizzical looks.

Donna stopped at a bush bearing half a dozen bright yellow flowers. Each had four primly arranged petals. She pointed to an unopened bud on the bush. "Watch. Watch. It's going to open." Our puzzlement over Donna's behavior was quickly replaced by curiosity about the primrose. We could see the pale yellow bud quivering on the branch. It began to unfurl. Within five seconds the bud had spun open completely. A perfect primrose flower bloomed on the bush. From bud to flower in seconds! The performance was most dramatic for a plant. Aside from watching a Venus's-flytrap capture a ladybug, I had never seen a plant do anything so remarkable. Type-A behavior; truly a plant in a hurry. A moment later Donna singled out another bud that performed the same way.

Animal behaviorists can be found in universities throughout the world; but if any botanists call themselves plant behaviorists, I am not aware of them. Were such a profession to exist, an expert might focus on carnivorous plants that reverse the food chain on animals, or some mimosa plants whose leaves close when touched, or plant species that catapult their seeds. In the potential field of plant behavior, the primrose performance would be a main event.

But one need not be a botanist to theorize about nature's mysteries. Anyone can contemplate nature and enjoy reaching conclusions without knowing what others think or know (or think they know). You can seek professional explanations later. So without books or botanists we began to speculate.

We assumed that these primroses were pollinated by nocturnal animals such as bats or moths. Some flowers pollinated by temperate-zone bats emit a smell of dead meat. Perhaps the smell of such flowers attracts carrion-loving insects, which some bats prey on. In the process of moving from one smorgasbord to another, maybe the bats pollinate the flowers that provide the meals. I'm not sure what the bugs would derive from this relationship. They would simply appear to be duped into attending a meal that is never served, except to the bats.

However, the scent emanating from the primrose was more like a perfume counter than an abattoir; we rejected the bat-pollination hypothesis in favor of pollination by moths. Perhaps the primrose's fragrance is similar to that emitted by certain female moths, a sure lure for the males. Maybe moths on a reproductive mission of their own were the pollinators of primroses. After all, insects are vital for the pollination of many plant species.

But why was the flowering act completed in such an abrupt fashion? Most flowers produce their sweet smells over several days. The evening primrose packages its scent into one sudden burst of fragrance. Perhaps the flower's perfume is propelled a greater distance into the night air and is, therefore, more likely to reach a waiting audience, such as hawkmoth pollinators. The flowers wilt the following day, but other buds await their turn the next evening.

I felt confident that at least some of our suppositions were correct. I returned to my sister's house with plans to look into the matter more thoroughly. As we sat down to our evening meal for the second time, we agreed that Donna's primrose spectacular had been well worth a belated supper.

The primrose experience reinforces how ignorant we can be about even the most common plants and animals. Ecologically, my observations of the evening primrose seem amateurish and naive, for botanists and horticulturists presumably have known of this phenomenon for years. But as far as I know, others have only observed the phenomenon, perhaps marveling as we did at the behavior of the primrose, without investigating further. Eventually, the primrose act will become the object of the insistent prying and testing of a scientist, of someone who has to understand exactly what is going on, and why. Someday we will know both how and why the evening primrose opens so abruptly.

Unsuspected traits such as the one observed in the evening primrose are a feature of every species. These traits are revealed only when someone is present at exactly the right time. Each year, research ecologists discover new attributes of the most common plants and animals, not to mention discoveries about rare and unusual species. This is especially true in the tropics, where species richness is the greatest—and our knowledge the weakest.

Certain species of tropical lilies serve as other outstanding examples of the breadth of unusual traits found in the plant world. Although not classified by botanists as a true lily, but as a member of the arum family that includes philodendron and caladium, the voodoo lily of Asia bears a beauty typical of many species in both groups and reaches a height of almost three feet. A garden lover would be stirred by its appearance. But voodoo lilies have one characteristic that would make them unpopular in the garden. Like some bat-pollinated species, they smell awful during the pollination period. An even more bizarre trait is that they heat up. In some instances the temperature inside a voodoo plant can reach 110 degrees Fahrenheit, even in cool shade.

Pretty plants and ones that smell bad are commonplace, but heat production is generally reserved for members of the animal kingdom. So how does one explain the phenomenon? When seeking a natural history explanation for a trait observed in a plant or animal, consider these things: How does the trait enhance feeding, protection, or reproduction in the species? These are the three essentials for remaining alive and perpetuating one's genes, and they serve as good first guesses for why a plant or animal does what it does.

The thermogenic properties of the voodoo lily apparently serve to increase the strength and dispersal of the odors that attract pollinators to the lily. Many insects, such as some scarab beetles, are attracted to the smell of decaying meat. Chemicals in the voodoo lily produce such a smell. The increased temperature allows the smell to travel, thus luring insects from afar. In addition, the temperatures inside the plant keep the beetles and other pollinators warm and active, thus insuring maximal contact with the reproductive structures of the flowers.

The biochemical explanation of how plants raise their temperatures is complex. Thermogenic plants were discovered in the 1700s, but only recently have biochemists begun to understand the physiological processes. Briefly, the normal metabolic pathways are altered during reproduction so that instead of storing energy, the plants produce heat. The potential applications of plants' thermogenic properties are intriguing. Perhaps scientists can transfer these heat-producing genes to other plants and modify their metabolism. Certain agricultural crops might be made more resistant to freezing or cold storage. Perhaps the metabolic pathways in pest species could be altered so that heat production is substituted for energy storage. Unable to nourish itself, the pest plant would die.

One other attribute of the voodoo lily has fascinating medical possibilities. Although the details are far from understood, prior to producing heat, voodoo lilies produce salicylic acid, the primary pain-relieving agent in aspirin. With proper biochemical enterprise, we might all stand to profit medically from the voodoo lily.

Discovering unusual biological properties, especially in plants or insects, is more likely in the tropics because of the awesome species diversity there. Yet even North American (and other temperate-zone) species have their special acts. Perhaps the best-known native North American plant species with heat-producing properties is the eastern skunk cabbage. In the Northeast, skunk cabbages are among the earliest plants to emerge in the spring, often pushing up through a covering of snow melted by their heat. Some plants have been reported to raise their temperature 45 degrees Fahrenheit above that of their environment. Skunk cabbages, like voodoo lilies, attract (and are often pollinated by) dead-meat-loving insects, as anyone who has been in their vicinity might surmise from the smell. Who knows, perhaps the skunk cabbage will one day be found to have properties that can enhance the well-being of humans.

Environmentalists lobby and cajole and harangue ceaselessly about the need to preserve tropical rain forests. Rightly so. The diversity of species in rain forests is astronomical and the biologically bizarre is commonplace. But we should also learn to appreciate species native to the temperate zones, which are full of fascinating plants. Every plant species, wherever it is, follows a blueprint for living that is peculiar to it alone.

Take, for example, the feeding habits of a special group of North American plants. When an animal eats a plant, we take it for granted. Happens every day and seems natural. But let a plant eat an animal, and it's a special case. Nonetheless, throughout North America and much of the world, animals are being eaten alive by plants every day. The plant-eats-animal phenomenon was identified many years before Charles Darwin wrote *Insectivorous Plants* about the subject in 1875. Today, botanists refer to these plants as carnivorous rather than insectivorous, because not only insects but animals such as small birds and frogs are on the menu.

Plants that capture and digest animals occur in many parts of the world, and several distinct kinds live in North America. Pitcher plants and Venus's-flytraps are relatively well known; sundews and butterworts are less so. Bladderworts are probably the least familiar of our carnivorous plants. But relative notoriety aside, they all have one feature in common: They get a portion of their nutrients from animals they catch. Most of these plants live in highly acidic habitats low in soil nutrients— habitats that are not particularly dramatic in appearance. But to survive in such environments, plants must compensate for the low availability of nutrients. What better way than to have nutrients brought to you from the outside, packaged in the form of animals?

Carnivorous plants lure and capture their preys in different ways. Some plants exhibit almost as much movement and animation as carnivorous animals. The one with the most style is the showy Venus's-flytrap, whose modified leaves perform a special magic trick for insects. The two halves of the trap look like a large, opened butter bean with long spines around the edges. The scent from nectar glands on the inside of the open leaf attracts flies and other insects. When a bug alights and one of its legs hits any two of several hair triggers, the halves of the trap slam shut, fast. Too fast for a fly to escape. The flytrap then secretes digestive juices and over the next several days absorbs the insect.

Although less well known, bladderworts are equally fascinating in their method of prey capture. Their act is one of the most amazing, yet underpublicized, feats of a North American plant. These small, mostly aquatic plants float at the water's surface, armed with thousands of bladders, each one a special trapping device that offers an invitation no small, swimming creature can resist. Animals caught by bladderworts are the ones that swim too close to the hair triggers on the outside of the bladder trap. When they touch a trigger, a tiny door on the bladder trap snaps inward. The helpless victim is sucked inside with a rush of water caused by expansion of the trap. The door immediately slams shut again, and the digestive process begins. The door opens and shuts in less than a second! This is about as fast as plants can do anything, and because of this phenomenal speed, the operation has been difficult to photograph. For many years botanists were unable to understand the process. The largest bladder traps are about the size of a match head, so the bladderworts' preys are mostly small insects and protozoans. These tiny carnivorous plants are merciless with mosquito larvae, which squirm around so much that if they are in the vicinity of a bladderwort, they invariably hit a trigger.

Pitcher plants use their stalk columns as highly effective pitfall traps. In some species the column is only a few inches high, but it can be almost three feet tall in the yellow trumpet. With downward-pointing hairs around the lip of the column and digestive liquor at the bottom of the flask, pitcher plants mean certain death to many insects. The bug that takes a misstep over the edge of the tube will soon become part of the plant world as it is digested and absorbed. Sundews and butterworts use a sticky trap to capture preys. They may smell good to an insect, or perhaps the glistening coating of mucilage has the appearance of nectar. An insect landing on the leaves quickly discovers that its feet are glued to the plant. Digestion occurs right there on the surface of the leaf, although some of the sundew plants have tiny tendrils that slowly close around the captive insect to ensure it stays for dinner.

In one sense, carnivorous plants are no different from other plants; they simply have evolved mechanisms to get nutrients through a nonsoil route. They have capitalized on a system suitable for plants restricted to wetland habitats with high acidity and low nutrients. An often-observed ecological phenomenon is that organisms with specializations for an extreme habitat are unable to thrive in more normal habitats. One explanation is that the species achieves an advantage over other species because of its specialization but does not have the genetic latitude to compete with these other species under normal conditions.

But competition in the plant world—as elsewhere—can take some unusual paths. Another group of highly specialized North American plants have flowers pollinated by insects, have seeds transported by birds, and take their water and minerals from trees rather than from the soil. With the additional clue that the plants are closely associated with Christmas, one can soon guess that mistletoe is the plant in question. *Mistletoe* refers to any of more than two hundred species of parasitic shrubs found worldwide. Varieties are found on every continent except Antarctica. Mistletoe occurs throughout the southern United States, from the Atlantic coast to California.

Having true parasitic properties, mistletoe is devoid of roots. Instead, the dark green shrub anchors itself with rootlike extensions (a difference significant, perhaps, only to a botanist) that suck water and nutrients from the host tree. Unlike Spanish moss, which extracts water and nutrients from the atmosphere and uses a tree, dead or alive, only for support, mistletoe requires a living tree for survival. In the southern United States, tiny yellow flowers bloom on the evergreen mistletoe from fall into winter. The familiar white berries begin to form after pollination and resemble little packets of glue around tiny indigestible seeds. A mistletoe plant can be either male or female, and like a holly tree, only the female plant has berries. Some reports advise against eating mistletoe berries, as they could be lethal to humans. Birds, however, seem to be immune to this particular toxicity.

The birds are essential to the welfare of the mistletoe plants. Their dispersal and propagation depend largely on birds, which eat the berries but do not digest the seeds. Recent studies suggest that seeds are most likely to survive and grow if a bird regurgitates or otherwise deposits them on the same species of tree on which the parent plant lived. Still, the migration of a flock of cedar waxwings can result in a newly developing mistletoe plant being a long, long way from its parents' home.

Mistletoe thrives where birds perch, which is why the plants are most often found in the uppermost branches of tall trees. In the southern states, mistletoe is commonly associated with certain large trees, particularly oaks. Its usual absence from southern pine trees and evergreen hardwoods such as magnolias may be due to the need for direct sunlight during the flowering period. Mistletoe's parasitic lifestyle is unusual among flowering plants. The effort to obtain water and minerals and even space itself is intense and highly competitive among most plants. But mistletoe does not encounter such problems. Tree limbs, a ready source of water and minerals for this unusual little plant, are available throughout the South. If mistletoe is absent from the uppermost branches of a tall oak, the explanation may simply be that no bird has dropped a seed there.

Perhaps in part because of its many unusual ecological properties, mistletoe has been singled out over the ages as possessing unique qualities. In Scandinavian legend, mistletoe was the only organism in the world from which Baldur, a son of Odin, was not protected. Thus, a dart made from the seemingly insignificant mistletoe was the cause of Baldur's death. Mistletoe is also associated with Druids, the mysterious, oak-worshiping sect that inhabited the British Isles centuries ago. To the Druids, mistletoe was considered a plant of honor and power. According to legend, when the parasitic plant was found growing in an oak tree, the Druids performed mystical sacrificial ceremonies at the tree on the sixth day after a full moon. The Druids reportedly used a golden sickle to harvest mistletoe from sacred oaks.

Mistletoe was used as a romantic lure in England at least as early as the 1500s. In 1520, William Irving wrote that a young man should pluck a berry each time he kissed a young girl beneath the mistletoe. A version of this tradition persists today in secular Christmas decorations. But, although the berries appear just in time for Christmas, mistletoe is not used in churches. Maybe mistletoe is excluded from today's religious decorations because of its association with the Druids.

Next holiday season when you're standing under the mistletoe waiting for a special someone to approach, take a moment to reflect on the history and the ecological wonders of this fascinating plant. For that matter, consider that any plant you look at in your yard, garden, or city park, or in the wild may have ecological secrets yet to be revealed because the plant has not been observed by the right person at the right time. We are surrounded by a natural world full of ecological intrigue and mystery, all wrapped up in the plants and animals around us.

Disguises of a Caterpillar

Trip Lamb and I had taken our herpetology class on a field trip to observe reptiles and amphibians at the U.S. Department of Energy's Savannah River Site in South Carolina, one of the largest protected wildlife habitats in the eastern United States. During a day and part of a night, the class caught, saw, or heard forty different kinds of reptiles and amphibians, more than are found in most states.

But it was a species of insect, a beetle, I remember most vividly from that night. We had gone half a mile into a swamp where the water was knee-deep. The class had headed back, but I lingered behind, alone, to listen for tree frogs. I turned off my flashlight and was greeted by a spectacular light show. Above the black swamp water, amid a forest of cypress and tupelo gum trees, the darkness glimmered with the twinkling lights from dozens of fireflies. Each light was reflected in the standing water, doubling the visual impact of these living sparks. It was the most dramatic display of these cold lights of the insect world I had ever seen. As a biologist, my first question was, What are they doing out here?

Lightning bugs belong to a family of beetles known by the scientific name Lampyridae, a suitable name for a bug that carries a lamp. By the latest counts, about two hundred species of the family are found in North America, and nearly two thousand are found throughout the world. The majority occur in the tropics, and each species has its own intricate, highly specialized way of life.

In virtually all of the common species seen in backyards across the United

States, only the males fly. They blink their lights in a code that advertises they are available for courtship. To attract the male, the female firefly, who also has a light, returns the signal from her location on the ground or vegetation. However, because many species are often active at the same time and place, the codes vary, thus preventing mating mixups. Next time you are outside on a summer's evening, look for differences in the rates and patterns of blinking by individuals. If the repeated patterns vary, the fireflies are quite possibly different species.

In some species, the females also fly and presumably signal differently from the males, to avoid ending up with the wrong sex when on a courtship mission. The ecology of most species of fireflies is poorly understood because of the complexity of the mating system. Also, some species have a limited geographic distribution, and some are rare wherever they live. However, James Lloyd of the University of Florida, through years of detailed observations and experiments, has revealed a remarkable level of behavioral intricacy among these lanterns of the night.

Deception is a common trait among lightning bugs. This is especially true in tropical situations, and possibly any situation in which several different kinds occur in the same area. A compelling reason to be deceptive is that some female fireflies as well as the larvae (which also have lights) eat other species of fireflies. When the male of another species is seen in the night sky, the impostor changes its own coded flashes to mimic the female of the same species as the flying male. When the male flies down, ready for courtship, he finds instead that he is expected for dinner—as the main course.

Some species may have evolved an additional step in this complicated process. In these species, it is suspected that when a male gets a blinking "okay" signal from the ground, he flashes a false code before flying in to mate. If the female or larva answers the false code, the male will have discovered the trap and keep on flying in search of a mate of his own species. If a different code evokes no response, the male tries the code of his own species again for confirmation.

As for why some fireflies seek mates in the middle of a flooded swamp, I still have not found the answer. One of the bewitching aspects of ecology is that quite possibly no one knows what the answer is—yet. I never cease to marvel at how complicated the ecology of some, perhaps all, animals really is. Even some of the simplest species enjoy a fascinating complexity in life, often with wonders hitherto hidden from observation. Some recent findings about invertebrates should instill in each of us an appreciation of what modern scientists have discovered.

Erick Greene of Princeton University studied a type of moth that lives in southern Arizona. The larvae, or caterpillars, of this particular species had never been described before. Greene discovered that the moths live on oak trees and have two life cycles during a year. Some adults become active in late winter and produce eggs that hatch into larvae in spring. Others lay eggs in summer and the larvae hatch soon after. So far, this may not sound particularly unusual, certainly not that different from a lot of other insects. What the hatchling caterpillars do at the two different times of year is where the wonder comes in.

Larvae that hatch in the spring feed on the male flowers of oak trees. These fuzzy, yellow oak flowers, or catkins, are familiar to anyone who has taken a careful look at an oak tree in the spring. Not only do the moth caterpillars eat the catkins, they look almost identical to them. Thus, when a caterpillar is having a meal, it resembles the food it is eating. Obviously, this is great camouflage for the caterpillar, which could become a tasty morsel for a bird if it were discovered.

What a fantastic system for providing food for young caterpillars and at the same time protecting them from predators. But what about the larvae that hatch in the summertime, when all the catkins are gone? Sure enough, this species of moth has another strategy. The summertime caterpillars eat leaves instead of catkins. And they quickly assume the shape, texture, and brown color of a tiny oak twig.

The larvae from the two different seasons do not vary genetically, so an obvious ecological question is, How does each caterpillar know to look like a catkin or a twig? The investigator experimented with raising caterpillars to find out. He captured some female moths that were ready to lay eggs in April, others that would lay eggs in July. He also collected and froze catkins and leaves of oak trees from the two different times of year. Then, when eggs hatched, he was able to feed half of the larvae one diet and half the other.

As it turned out, the season the caterpillars were born was not the determining factor. Their future appearance was dictated by what they ate. Caterpillars fed catkins began to look like catkins, even in summer, when catkins would not normally be present. Those fed leaves began to resemble twigs. Further experiments and chemical testing suggested that caterpillars were responding to particular compounds, called tannins, that are present in the oak leaves but not in the catkins. Larvae that eat tannins assume the shape of a twig. If the tannins are absent from the diet, the larvae develop into the catkin type of caterpillar.

While some insects go to great lengths to avoid being eaten, others excel in ways to capture what they want to eat. And the manner in which they capture their prey varies widely. Bolas spiders, for example, do not build typical silk spiderwebs. Instead, they produce a type of glue and then hurl a sticky line of the glue at their insect prey. The technique works somewhat like the ropes with weights on the end, or *bolas,* once used by South American cowboys. In both cases the quarry is snared by the line and captured.

The discovery of bolas-twirling, or sticky yo-yo, spiders was fascinating, but the true mystery for ecologists arose when it became apparent that all the bolas spiders they examined ate only male moths. Why not females, too? The answer to that ecological question was discovered through chemistry. Pheromones are chemicals released by animals to signal other members of the species. A carefully planned study with bolas spiders, conducted by Mark K. Stowe of Harvard University and his colleagues James H. Tumlinson and Robert R. Heath, revealed that the spiders produce chemicals that mimic pheromones used by female moths to attract males prior to mating. Each species of moth produces a different pheromone. To ensure a wide selection of mealtime choices, the spiders fill the air with a variety of perfumes that include the critical ingredients guaranteed to attract male moths of numerous species.

But the spiders' ingenuity can be matched by the preys' resourcefulness. Consider the special relationship between snowberry flies and zebra spiders. These jumping spiders do not build webs; they must catch their prey face to face. As the name implies, zebra spiders have stripes, and they will eat a snowberry fly in a heartbeat if given the chance to pounce on it. However, the snowberry fly has wings with stripes that look like the zebra spider's legs, and it knows how to use them when confronted. By raising its wings and assuming a particular posture a snowberry fly can make itself look like a zebra spider. Based on laboratory experiments by Monica H. Mather and Bernard D. Roitberg of Simon Fraser University, the ruse works well enough to save the lives of snowberry flies on a regular basis. The flies are presumably not dismayed at the prospect of being mistaken for a potential mate instead of a meal.

As if such chicanery in the bug world were not enough, a case of female impersonation among scorpion flies, harmless wasplike insects, has been documented by Randy Thornhill of the University of New Mexico. Scorpion flies begin courtship with the male inviting a willing female to dinner. The meal consists of an insect captured by the male and brought to the female, who accepts the offer to share a meal. However, fine dining is not foremost on the male's mind, and while the female is occupied with eating, the male mates with her. A male scorpion fly may seem to have an easy life, but in this insect's world getting a meal can be very costly. For one thing it takes a lot of energy. Also, in its habitat, an insect-chasing scorpion fly runs the risk of being trapped by web-building spiders. Some males have turned to transvestism as a less risky means of preparing for courtship. Landing beside another male with a recently captured insect, the impersonator lowers his wings in a manner characteristic of females. The first male, who assumes the other is a female, willingly hands over his prey, anticipating sexual favors. The deceiver scorpion fly takes the prey and quickly flies away. Within minutes he uses the free meal to entice the female of his choice.

Admittedly, all the examples given above occur in warm climates. But invertebrates, like all other organisms, are complex wherever they occur, including the Antarctic seas. The evolutionary contest between hunter and hunted occurs everywhere, and innocent bystanders can sometimes find themselves involved.

James B. McClintock of the University of Alabama at Birmingham and John Janssen of Loyola University in Chicago discovered a chemical association between two species of tiny marine invertebrates and a fish—an association that is as multifarious as relationships can get. One of the organisms in the relationship is an amphipod, a type of invertebrate related to sow bugs, or roly-polies. The amphipod is defenseless against fish that are major predators on small organisms in the region. In the clear waters of an Antarctic sea, the amphipod is easy prey. In contrast, another small invertebrate, known as a pteropod, is chemically noxious to the fish and therefore not a prey item. Pteropods are in essence tiny snails without shells, and they move through the water by flapping winglike appendages. Field studies based on dissections of fish have shown that amphipods are common prey for a variety of fish. But the bright orange pteropods are not eaten. A fish that grabs a pteropod will shake its head violently and spit the animal out.

But no tasty animal without any defense whatsoever would last long on the evolutionary playing field. And sure enough the amphipods have developed a defense that works—as long as pteropods are around. The investigators drilled holes in the sea ice and observed that many amphipods were carrying pteropods on their backs. Amphipods placed in aquaria were seen to pursue and grasp pteropods and actually hold the smaller animals captive. Using a scanning electron microscope, the investigators discovered that the amphipod uses pincers on its rear appendages to grip the pteropod so that it cannot escape. The amphipod then places the captured pteropod on its own back in full view and, still holding it with the pincers, goes about its business.

Presumably, then, amphipods capture pteropods as a chemical weapon against fish predation. The scientists ran laboratory tests to find out for sure. The experiments revealed that an amphipod carrying a pteropod was almost immune to fish attack, whereas one without a pteropod was eaten readily. Chemical defense is common in prey-predator relationships. Some examples of animals that use it are skunks, scorpions, and stinging nettle. But those species all produce their own chemical warfare. The amphipod-pteropod relationship provides a rare example of how a prey species actively exploits another species for chemical protection.

The amphipod pays a price for carrying around its kidnapped bundle of protection. With the added weight, swimming through the Antarctic sea is a bit more difficult. But not having to worry about every fish that swims by is apparently worth it. What does the pteropod get out of this relationship? Nothing, so far as the scientists have been able to tell. Although no pteropods were observed to die while held captive, presumably they do not maintain a normal diet. Certainly, being dragged around on another animal's back for a week or more would not seem to be in a pteropod's best interest. Ironically, the trait that spares it from being eaten by a fish is the very trait that results in its being abducted by another animal.

The interaction of individual creatures with other individuals and their environment is captivating. Entire populations acting as programmed biological units can be equally intriguing. A claim has been made by Nigel R. Franks of the University of Bath in England that "army ants form the most cohesive societies on earth." Statements such as this, about extremes in the world of living organisms, can be dangerous. Nonetheless, on the basis of Dr. Franks's research, I stand convinced of the truth of his claim.

Bees and many colonial ants and wasps function as individuals in the best interest of the colony. An army ant colony, however, operates like a single organism, a problem-solving unit. Together the ants operate day and night as a unit programmed to follow a plan of action, responding to special situations with a collective intelligence. Yet an army ant by itself will walk in circles until it dies.

Hundreds of species of army ants exist, primarily in the tropics, but the best known ecologically are the African driver ant and two tropical American forms. The African driver ant lives in colonies of more than twenty million ants. In one of the tropical New World species, the army ants in a colony go on a raid each day, all walking in the same direction. Movies have exaggerated the danger to humans, but it certainly does not pay to be an insect, large or small, in the path of army ants on the march. The swarm moves at a rate of less than one foot per minute, and few small animals that remain in place are overlooked or go uneaten. Movement of the ants through the jungle ground litter stirs up enough activity that birds, attracted to the scene, prey on mobile insects trying to escape the creeping army of ants.

Army ants are nomads with no permanent nest. Indeed, the nest where the single queen and developing young stay is actually alive, formed by the legs and bodies of thousands of ants that somehow regulate the temperature of the ever-moving nest. An average queen is estimated to live about six years, walk more than forty miles, and lay more than six million eggs. An army ant worker is in one of four castes, each differing in appearance and function.

The largest workers are called majors. A major looks like a pair of ice tongs with an ant attached. Majors close their huge pincers on anything they identify as an enemy of the ant colony. But the pincers work only one way—to close. This fearless defender of the colony never releases its hold after attacking; it gives up its life in a true death grip. Submajors, also large ants, make up a small proportion of the work force. They carry dead prey back to the nest and organize teams of other workers if the item is too heavy. The army ant transport teams carry food faster and more efficiently than do ants of most other species. One unusual feature of the transport process is that all teams carrying prey move at exactly the same rate. This is accomplished by their sophisticated social interaction. If an item is moving slower than the standard of the colony, the closest ant will join the transport team to help it move faster.

Most individuals are the so-called medium workers. These can and will do anything required. One intriguing feat is seen when army ants face an obstacle such as water or a crater. The medium workers form bridges, made by the bodies of hand-holding ants, that the others can walk on. The minims are tiny worker ants that serve as nursemaids to the brood in the nest. Because of the petite size and slow gait of the minims, the submajors often carry them during the migrations from one nesting area to another.

Because army ants ravage entire areas where they forage, they must choose a different direction each day to assure a plentiful food supply. A species studied in Panama was extremely predictable. Raids were made in straight lines at evenly spaced angles, like spokes of a wheel, so that overlap between raids was minimal. A colony ordinarily made fourteen raids from a nest site. On the fifteenth raid, the colony moved away from the nest site, traveling overland for fifteen days. They carried the young and created a temporary nest each day.

Many mysteries remain about army ants. For example, how do they know which direction to go each day, since no individual seems to be in charge? And what process does the colony of workers use to decide which of the six queens produced each year will be the ruler? We don't need Hollywood's early presentation of army ants as marauding, relentless, human flesh eaters to have a fascinating story. The real army ant is story enough, and there is a lot more of the story still to be learned.

Other ants have entered into a rather bizarre alliance with another type of insect, aphids. As sap-sucking residents that damage our shrubbery, aphids are hardly our favorite animals. But we are not the only ones out to get these tiny, soft-bodied insects. Aphids are prey for numerous other garden inhabitants. And being more or less defenseless themselves, they can use a little protection. Some species of ants actually fulfill the role of protector of the aphids. Stranger still is the discovery by Thomas Eisner and his colleagues at Cornell University that a camouflaged intruder may live within the aphid herd unnoticed by the shepherding ants. The intruder, the larva of a green lacewing, has found a way to go undetected by the ants while at the same time indulging a very selective diet—aphids.

The ant-aphid partnership has been observed in several different combinations of ant and aphid species. Aphids have sucking mouthparts to remove juices from the soft parts of plants upon which they live (often your camellias or boxwoods). The accumulation of plant sugars causes the aphids to excrete a saccharin-sweet material known as honeydew. Ants apparently find honeydew an irresistible food source and remove the substance without harming the aphids. When an ant colony becomes caretaker for the aphids inhabiting a particular shrub or tree, mutual accord is reached between the two species. The relationship is reminiscent of that between shepherds and their sheep. Although each individual sentry must periodically leave the bush and return to the ant colony located in the ground, a few guards always remain in the shrubbery with the aphids. The intricacies of the changing of the ant guard in an aphid herd have not yet been determined, but the ants have obviously worked it out to their satisfaction.

The green lacewing is a flying insect that lays its eggs on the leaves of alder trees, always in the vicinity of a colony of woolly alder aphids. When the eggs hatch, the aphid-sized larvae move into the colony. The lacewing larva is a voracious eater of aphids. It uses its hollow, sharp-pointed pincers to grab an aphid and suck it dry in less time than it takes you to drink a milk shake. However, a lacewing larva discovered by one of the ants is in trouble. The ant guards attack the aphid killer, and the first ant on the scene has the honor of dispensing with the unwanted guest. The ant bites the larva, lifts it up in the air, and hurls the helpless intruder over the side of the limb, banishing it from the bush forever.

But the discovery of lacewing intruders by ants guarding aphid herds happens far less frequently than might be expected. The reason: The lacewing has a camouflage strategy that relies on material from the body of the aphids. Woolly alder aphids look white because of heavy body secretions of a white wax. The exact function of this wax is unknown, but whatever its purpose, aphids moving along a dark alder limb look like a tiny herd of sheep.

When a lacewing larva enters an aphid colony, it has two goals: to feed on the aphids and to protect itself from ants. The second goal is of immediate concern in a heavily patrolled aphid herd. Consequently, the lacewing's first action upon encountering an aphid is not to eat it but to disrobe it. Research has determined that the larva uses its pincers to strip the white wax from the surprised aphid. Then, through a series of body maneuvers, it places the material on its own back. Enter the wolf in sheep's clothing. The costume of stolen wax is effective in fooling the ants, who apparently mistake the sheep-clothed lacewings for aphids. If a researcher blows the lacewing's cover by removing the wax, the ant patrol promptly throws the trickster off the limb.

Because of their small size, insects and other invertebrates lend themselves to observation and experimental manipulation that reveal a fascinating world of ecology and behavior. Since most animals are insects, the revelation of such complexity and intricacy among common insects should make us consider. What other natural wonders and mysteries exist all around us? And how many will we never discover because we have destroyed a delicate relationship before we even realized it existed? As we continue to find out about such intricate schemes of ecology and evolution,

23

we begin to realize how much we still do not know. With greater understanding of ecosystems and all their interconnecting parts, we can better appreciate our natural environments. Perhaps we will even learn to deliberate carefully before we tamper with them.

We need not look solely to outer space for ways to explore the mysteries of the universe. Plenty of opportunities for discovery are right here on Earth. And a small step by each of us may prove to be a giant leap for mankind in understanding the environment we live in.

The Advantage of Being a Cold Fish

In 1965, I stared into icy waters looking for life. Any kind of life, but especially reptilian. I was a student at Michigan State University, conducting my dissertation research at the W. K. Kellogg Biological Station. As one who wanted to work with reptiles, I found Michigan winters to be excessive, especially in length. Of course, being a herpetologist is not a prerequisite for having this attitude, but a herpetologist is likely to find little of interest amid ice and snow.

Painted turtles, like other reptiles, disappear during the winter in Michigan, from about November to April. By early April of that year, the winter had gone on long enough to suit me. I wanted to see a lake without ice. George H. Lauff, director of the biological station, had arranged for me to collect turtles during the warm months on the lakes at the Kellogg Bird Sanctuary. April is not noted for being a warm month in Michigan; nevertheless, a phone call from Joe Johnson at the bird sanctuary sent me out to one of the lakes. Joe said there was open water around the boat dock at Wintergreen Lake and that he had seen a turtle swimming beneath the ice shelf.

Well, this was back when herpetologists "knew" that turtles were not active in cold weather. Three earlier anecdotal observations and reports, by Archie Carr, Fred Cagle, and Owen Sexton, of turtles swimming under ice were assumed to be exceptional and unusual. Therefore, reptile ecologists did not waste time looking for turtles in half-frozen lakes. But the day was sunny, in fact brilliant, and by my

Alabama standards the winter had overstayed its welcome by several weeks. I was anxious to see my first reptile of the year; even a single, disoriented painted turtle would be a welcome sight.

I arrived at Wintergreen Lake, and Joe was right about the open water at the dock. Icy cold, clear Michigan water surrounded the wooden dock and extended for a hundred feet in either direction along the shore. The open water stopped at the edge of the ice shelf that covered the remainder of the lake. I was disappointed. The lake was not really accessible; only a few hundred square feet around the dock were unfrozen, far less than an acre of the twenty-acre lake. Still, the sky and water were clear and I had no other herpetological prospects. I decided to take a look. Standing in the bow of a cold aluminum boat, I pushed out into the water and used one of the oars as a pole to push along the shoreline. Within seconds of leaving the dock, I saw a painted turtle swimming along the bottom of the lake. I used my dip net, turning my first reptile sighting of the year into my first capture: a male painted turtle, its black shell margined with crimson, and black head and legs adorned with bright yellow stripes. A handsome creature, and very active—amazingly active for a reptile in ice water. I assumed this to be the turtle Joe had seen, an aberrant animal as anxious to rush the season as I was.

While I was putting the turtle into a collecting bag, the boat drifted farther from the dock, toward the ice shelf near the lake side of the open water. I peered into the crystal clear waters, hoping to see some sign of life, perhaps a fish, in the pristine habitat. The ice at the edge of the shelf was like a glass counter. And I saw life. Turtles. Dozens of painted turtles. Crawling. Moving. Swimming along the bottom at the edge of the shelf.

The observation resulted in my first technical publication in a scientific journal. My interpretation of the phenomenon I had stumbled on was not particularly profound. I concluded that the turtles were trying to raise their body temperatures on what was probably their first view of the sun in months. Subsequent observations of turtles in cold water, during periods when they were once assumed to be dormant, have been made by others. Perhaps a primary reason for trying to get warm enough to be active when ice is still on the lake is related to the breeding season of painted turtles; but herpetologists are still trying to resolve exactly when painted turtles mate, so maybe there are other reasons. Whatever the case, the painted turtles had taken advantage of the opportunity to raise their body temperatures; they were active under near-freezing conditions.

Temperature is one of the most obvious environmental factors influencing the ecology of plants and animals. Birds fly south to avoid winter. Reptiles become active when the ground warms up in the spring. Some tulip bulbs require a period of cold before they can sprout. And the fate of a peach or orange crop can be decided by a few degrees at critical periods. Geographic distributions of numerous plant and animal species are believed to hinge on when or how often the freezing point is reached during the year. But differences in temperature far above freezing can also be extremely influential in some situations. For example, some desert lizards remain dormant at 80 degrees Fahrenheit but become active when temperatures reach 90. At the cold end of the spectrum, biological mysteries are discovered each year. Some are solved. Some remain mysteries. The phenomenon of turtles under ice, observed almost three decades ago, still has not been fully explained.

Mammals, being warm-blooded animals, or endotherms, regulate their body temperatures internally. As outside temperatures drop, they must burn more fuel, or calories, to keep their body temperatures high. Food provides the energy that keeps an animal operating. Indeed, most of the food energy taken in by mammals is used to keep high, constant body temperatures.

Wintertime can be harsh for a mammal for two reasons. First, the cold itself means more energy is required to maintain body temperature at a constant level. Second, many of the food sources of the warmer seasons disappear. Whereas many humans have a problem with taking in too many calories, most animals spend their lives trying to get enough. Winter makes this especially difficult.

Seed and vegetation eaters, such as mice, usually do not emerge from cold northern winter homes until food is available in the form of buds and young grass shoots. Badgers, on the other hand, eat other animals and are active intermittently during the winter, searching for hibernating rodents.

Hibernation is a strategy used by various mammals during winter. But the approach varies with different species, depending on body size, sex, and feeding mode. Alan R. French of the State University of New York at Binghamton examined hibernation in several mammal species ranging in size from half a ton (bears) to only a few ounces (bats). A general biological rule is that larger-bodied animals can store more fat yet use less energy per pound than smaller-bodied ones.

Bats, on the other hand, actually lower their body temperature several degrees during winter. Their small size does not permit them to store enough fat to see them through the winter at normal body temperatures. By dropping their body temperature, they can conserve energy during winter dormancy. But this technique has a disadvantage. A cold, torpid mammal is effectively unconscious and cannot escape danger if disturbed.

Mammals that lower their body temperatures to survive winter do so at intervals interspersed with short periods of normal body temperature. Many of the western ground squirrels go underground and sleep for days at a time, but occasionally their body temperature warms up to normal. This gives the ground squirrel an opportunity to assess conditions and then retire again if winter weather persists. Larger mammals, such as woodchucks or groundhogs, have larger fat stores and maintain normal body temperatures for longer periods during hibernation. They also emerge from hibernation earlier. Their stored calories are more likely to see them through a long period of searching for food.

A bear can store enough fat during autumn to carry it through the winter. In a sense, bears and other hibernators stockpile energy when fuel is cheap. They seek out a cave or other protected area where they can nap during winter to avoid using up fat energy walking around looking for food when little is available. Bears maintain body temperature at the normal level and doze, an activity that does not exhaust stored calories. Thus they also reduce the rate at which the stockpile is used.

Developing an effective method for coping with winter is critical for survival, but it won't perpetuate the species. One of the most important things an animal does during its life is reproduce. It is not surprising that a relationship exists between the hibernation style and the reproductive strategy of a species. Most species mate in the spring, and adult males appear first, awaiting the females. The nonreproducing juveniles emerge last of all, when food is sure to be available, since their main interest in becoming active is to obtain a meal.

Many bears have a different strategy. They mate in the fall and the female goes through gestation and has her cubs while she hibernates. When the female bear reappears on the scene, her cubs are with her, looking for their first solid meal.

The reproductive strategy of hibernating bats may be the most unusual of all. Because of the uncertainty of when insects will be available to replenish their meager fat stores, bats do not have a spring mating season. Instead, they mate in the fall. The female, unlike most other mammals, retains the male's sperm but does not fertilize the egg. She waits until spring temperatures are warm enough for insects to appear, emerges from hibernation, and then allows fertilization to occur. Not surprisingly, bats are one of the few mammals in which the females become active in spring before the males. The first bats you see in the spring are probably soon-to-be mothers carrying their unborn babies.

We ecologists often understand the natural history of one species in detail or a particular phenomenon in a general way. But we must not become complacent, believing that we know it all. Each species differs from all others, often in ways we may not understand. For example, not all bears or all bats handle the winter as described above. The behavior patterns and ecology of each species have been synchronized to particular environments over millions of years.

Winter conditions in the southern United States may be ideal for warm-blooded species, but southern winters can still get too cold for most reptiles. The American alligator would make the top ten list of animals characteristic of the South. Where does this air-breathing, water-dwelling reptile spend the winter? What happens to American alligators during winter in the northern parts of their range such as the Carolinas, where ice forms on the surface of lakes and ponds during especially cold winters? Naturalists assumed that alligators retired to dens under the bank or to the bottom, where they remained throughout the cold period. With their respiration rate lowered by the cold water, breathing air was thought not to be a critical issue. If necessary, the alligator just drifted up to the surface and took a deep breath through its nose. But what would happen if a sheet of ice prevented the alligator's snout from reaching the air?

As it turns out, alligators are not passive creatures that let a cold winter turn them into inactive, sleeping logs. They actually fight back against the cold with some fascinating tricks. As water temperatures approach freezing, the alligator moves to shallow water, keeps its nose above the surface, and waits for the water to freeze. If freezing has already begun, it may nudge its heavy nose against the thin, newly formed ice to break open an air hole, in the manner of a bowhead whale in the Arctic. With its body extending down into near-freezing water, the cold-blooded, air-breathing reptile remains with its nose protruding through the ice as long as subfreezing temperatures exist.

This remarkable phenomenon had gone unrecognized by scientists until the 1980s, when two research groups independently revealed this behavior. Their findings about the behavior of big alligators were identical. During the unusually cold winter of 1981, the two different groups happened to be following the movements of adult alligators with the use of radio transmitters. One research program was conducted in the Croatan National Forest in North Carolina. John M. Hagan, Paul C. Smithson, and Phillip D. Doerr of North Carolina State University were the investigators. The other research was conducted by Larry D. Vangilder, Richard T. Hoppe, and Robert A. Kennamer of the Savannah River Ecology Laboratory in South Carolina. This hitherto unsuspected finding about the winter behavior of alligators makes a worthwhile point: A lot is still to be learned about even the most obvious of animals.

One remaining mystery is, What happens to young alligators too small to break through an ice covering? How do they breathe through a cold winter? In an attempt to find out, we watched a three-foot animal under a sheet of two-inch-thick ice that covered an experimental pond at the Savannah River Ecology Laboratory. The little alligator made numerous nose thrusts against the ice layer but was never able to break through. We finally knocked a hole in the ice with a boat paddle and came back later to find its small snout at the open spot, completely surrounded by ice but in contact with the required air. What if we had not been there? Do the young usually stay close to an adult and share the air hole the big one makes? Or is this simply one explanation for why alligators are a southern phenomenon, living in places where ice seldom forms?

Factors such as a shortened growing season and a limited period of warmth for egg incubation might limit the northward extension of alligators. Perhaps in more-northern habitats the ice covering becomes so thick that even some adults cannot break through. The addition of several inches of snow on top of the ice would definitely hamper the breathing of a stuck-in-the-ice alligator. The coldest places where alligators naturally occur do not typically experience heavy snow along with thick ice.

A fascinating feature of animals is that certain species manage to survive in environments we define as uninhabitable. Deserts, polar oceans, and the perpetual darkness two miles beneath the sea are all extreme environments from the human standpoint. Yet some species actually thrive under these seemingly harsh conditions. Clearly, they must have adaptations that permit them to do so, and an understanding of what these are can reveal the remarkable capabilities of some species.

Certain marine fish, for example, live in the Arctic and Antarctic seas, even below the polar ice caps. In this environment the water is not merely cold; the fish survive in temperatures that may be several degrees below freezing. The reason for this is that saltwater has a lower freezing point than freshwater. Thus, a polar sea may still be liquid at 29 degrees Fahrenheit, 3 degrees below the freezing point of freshwater.

Some fishes produce antifreeze in the blood that prevents them from becoming blocks of ice. In general, different families of fishes use antifreezes that are chemically quite different. Studies conducted by James A. Raymond of the Alaska Department of Fish and Game and two colleagues, Peter Wilson and Arthur L. DeVries, determined how these antifreezes interact with ice. They found that each compound works in essentially the same manner, resisting freezing by binding to the same specific crystal faces of ice in the blood at subfreezing temperatures. Why only certain crystal faces of ice are so special remains to be discovered.

Although most land animals deal with extremely low temperatures by seeking a warmer place during winter, at least one kind of turtle has a different solution—just get cold and freeze. All turtles lay their eggs on land, usually by digging a hole in dirt or sand, depositing the entire clutch, and then covering the nest. Most turtle eggs hatch in the autumn, but some hatchlings do not leave their shallow nests until the following spring. Thus, in Canada and the northern United States, painted turtle hatchlings are entombed only a few inches beneath the soil for the winter. They must sometimes endure soil temperatures as low as 25 degrees.

Justin Congdon of the Savannah River Ecology Laboratory, during the course of studies with freshwater turtles on the University of Michigan's E. S. George Reserve, discovered that hatchling turtles that spend the winter in the nest differ in body composition from those that leave the nest during late summer or fall. He found that the eggs of overwintering hatchlings have proportionally more lipids, which may be in the form of fat or oils, than do the eggs of those that leave the nest early. More lipids mean more energy. Therefore, the baby turtles can go from late summer to the following spring on their own fat reserves without eating. He suggested further that hatchling turtles may even be capable of producing antifreeze compounds. Experiments, by Kenneth B. Storey and Janet M. Storey of Carleton University and their colleagues, disclosed that hatchling painted turtles exposed to subfreezing temperatures produce significantly higher levels of glucose, other carbohydrate products, and amino acids in the blood than do those kept at normal temperatures. These may function as antifreeze products, but the mechanisms are unknown.

A more important discovery, however, was that the baby turtles can survive

31

when more than half of their internal body water freezes. The painted turtle is one of the highest vertebrate life-forms known in which the freezing of body fluids is tolerated during hibernation.

Arctic ground squirrels can endure the lowest temperatures known for any mammal during hibernation. Historically used by Eskimos for food and clothing, these ground squirrels of the Far North must hibernate more than seven months of each year to deal with the cold and lack of food. Arctic ground squirrels were observed by Brian M. Barnes of the University of Alaska to withstand deep body temperatures as low as 28 degrees without harm. The plasma of animals with temperatures below the freezing point (32 degrees) was normal, and no antifreeze compounds could be detected. Arctic ground squirrels have apparently acquired the ability to undergo prolonged supercooling, a below-freezing condition before ice crystallization occurs. The metabolic savings from lowered body temperatures is advantageous. By not having to maintain temperatures above freezing, as other hibernating mammals do, these squirrels can preserve energy stores during a long winter.

A discovery about reptile eggs has provided information about the consequences of even minor changes in what we would consider moderate temperatures. Among many groups of reptiles, sex determination, that is, whether an individual becomes a male or a female, depends in large part on temperature. The most dramatic example is provided by some of the turtles: Eggs incubated at temperatures above 86 degrees produce females. But if the eggs are kept at temperatures below 82 degrees, all the developing young will be males. Under normal conditions, the proportion of males and females (the sex ratio) is expected to be approximately 50:50 in most animal populations. Imagine the potential influence of the temperature phenomenon on the sex ratio in a population of turtles.

Richard C. Vogt of the University of Wisconsin found that map turtles living in a section of the Mississippi River were mostly females. Upon further study, he found that females nested on sandbars exposed to open sun. The eggs were incubated at high temperatures and produced mostly females. Males occurred occasionally, when mother turtles deposited their eggs in the few shaded areas. Another investigator reported that in recent years the sex ratio of green sea turtles in the waters around a particular island complex has changed dramatically. Although the sex ratio was once 50:50, there is now a preponderance of females and a scarcity of males. His explanation for the change is that resort development on the islands has resulted in the elimination of tree cover along the beaches. The only nesting areas remaining are those in open, unshaded areas. Thus the nests are warmer and most of the babies produced are females.

However, as is often true in biology, few rules are so consistent that we can confidently predict what will happen. And indeed, the American alligator, a reptile relative of the turtles, has completely confounded the issue of temperature-dependent sex determination. Whereas above-average warm temperatures in the nest produce *female* turtles, Mark W. J. Ferguson of the Queen's University of Belfast and Ted Joanen of the Rockefeller Wildlife Refuge in Louisiana found that such temperatures produce *male* alligators. Likewise, only female alligators are produced at cooler nest temperatures. Thus turtles and alligators are completely opposite in the way that temperature affects the sex ratio. In addition, the sex of some turtles is determined genetically, as is true of humans.

The temperature phenomenon is of scientific interest to ecologists and evolutionary biologists. It is also used to advantage by animal breeders who want to produce more of a particular sex. But one of the more fascinating considerations stems from a hypothesis, proposed by Ferguson and Joanen and furthered by James R. Spotila of Drexel University, that the influence of temperature on sex determination may have caused the extinction of some dinosaurs. The theory hinges on an assumption that the sex of some of these ancient reptiles depended on nest temperature. At the end of the Mesozoic era, major temperature changes occurred on a global scale. If average temperatures throughout the world rose or dropped several degrees over a short period of geologic time, some species may have begun producing young of only one sex. If so, whether male or female, the final result would be no more mating. And no more offspring.

Some scientists, however, do not think that a gradual change in the world's temperature over a few thousand years could have such an uncompromising impact on species. A study with a small coastal fish, the Atlantic silverside, challenges the notion that organisms cannot adjust their sex ratios in response to even relatively short-term environmental change. The sex ratio of Atlantic silversides becomes balanced, approaching the 50:50 ratio, even when it is initially skewed in favor of one sex or the other.

The determination of the sex of Atlantic silversides is intriguing in its own right. Although genetics has an influence, the sex of these fish can be determined to some degree by the temperature of the water during the larval period. Along the South Carolina coastline, silversides born during the cool temperatures of spring are predominately female. Those born during summer are mostly male. Because silversides are born during both spring and summer, over the course of a year approximately equal numbers of males and females are produced. However, the same species of silverside is found in Nova Scotia. At that latitude, temperature has no effect on sex determination. Instead, sex is determined genetically. In between South Carolina and Nova Scotia, along the New York coast, water temperature partially affects the sex ratio, but not as strongly as in South Carolina. That is, cold water still results in a shift toward more females, but the ratio is less skewed than in the South Carolina area. This suggests that an intermediate state exists between genetic and environmental influences. Taking advantage of this discovered trait in Atlantic silversides, David O. Conover and David A. Van Voorhees of the State University of New York at Stony Brook confirmed a theory proposed more than sixty years ago.

A 1:1 sex ratio is characteristic of most animal populations, with one female being born on average for every male. The theory, in simplest terms, is that if the sex ratio becomes significantly imbalanced from 1:1, natural selection will result in more of the less-abundant sex being born. Thus, if a population has twice as many females as males, then the average male fathers twice as many children as an average female produces as a mother. Therefore, an average male's genes would have a higher probability of surviving to the next generation than an average female's. In such a situation, it is "better" to be a male. If males were the more abundant sex, the reverse would be true. The genes in question are not those typical of sex chromosomes; they are genes that make it more likely that one sex will be produced than the other.

Although accepted by biologists, the theory had no convincing experimental evidence to prove that animal populations were even capable of making such a shift—until now. In laboratory experiments, the researchers raised thousands of Atlantic silversides at constant high or low temperatures, simulating those temperatures most likely to produce males or females in South Carolina and New York. After each generation, the scientists determined the number of each sex. Some of the experimental populations started with many more of one sex than the other. But, sure enough, after several generations each population had reached a sex ratio of 1:1, regardless of the water temperature.

Remember, if left at a constant temperature, silversides from South Carolina and New York should have an excess of one sex or the other. But the sex ratio gradually became balanced during the experiments, supporting the theory: Natural selection favors genes that produce the sex that occurs in lower proportion.

The findings with Atlantic silversides emphasize the subtlety of the response of organisms to their thermal environment. Such studies also confirm the complex and sensitive balance in which the earth's ecosystems rest, delicate relationships we may never fully understand. What might happen if we experience the global warming trend predicted by some climatologists, if we artificially raise the earth's temperature by several degrees in only a few decades? Sex ratios of reptiles and fish might adjust, but the temperature change could have a dramatic effect on other systems that have not evolved to make such adjustments. We do not really know how such impacts might affect the natural world—or even how they might affect us.

Findings about how organisms respond to temperature physiologically, ecologically, and evolutionarily may eventually be applied to problems that face humans. The use of cryogenics in medicine, the treatment of frostbite or heatstroke, even temporary alterations of the human body to withstand cold or heat are but a few possibilities.

All organisms are evolutionary products of the environmental history of their ancestors. Ecologists attempt to categorize the ecology of species living under different, definable environmental conditions. Comparative environmental categorizations include desert versus nondesert systems, winter versus summer patterns, and tropical versus temperate climate conditions. Temperature is one of the most easily measured environmental variables that clearly have an effect on the physiology and behavior of organisms. It is, however, merely one of the vast multitude of environmental variables to which organisms must respond. Moisture, light, and salinity are a few more.

Organisms use an array of survival strategies to live under a variety of environmental conditions. A more thorough knowledge of the life around us gives us an appreciation of the adaptability of the plants and animals that share our world. It should also make us acutely aware of their vulnerability. By preserving and exploring the natural world, we may even find answers for which the questions have not yet been asked. Research that enlightens us about such matters should be encouraged.

Why You Need a Rhinoceros
If You Own a Trewia Tree

Trip Lamb and Peter Stangel both made fun of my bird feeder; it attracted nothing but sparrows—English sparrows to me, house sparrows to the Audubon Society. Whatever you call them, my bird feeder was the fast-food outlet for dozens of those little brown birds, descendants of the sparrows brought to the United States from England in 1852, a true housewarming gift from overseas. Trip and Peter obviously regarded a bird feeder that lured only English sparrows as low-class and inferior.

Hoisted on the petard of mediocrity, I gave a predictable response. I pointed out the fascinating behavior of sparrows. The self-proclaimed dominant males puff up their outer feathers, lower their wings so that the tips drag the ground, and strut toward other sparrows in an aggressive display. Presumably such a performance allows one to get extra sunflower seeds. I indicated that even sparrows hold an ecological message for us and that my monotypic bird feeder could probably be revealing. Desperately I sought a worthwhile excuse for attracting only drab brown birds from another continent to my bird feeder.

Trip was present when the ecological message was delivered, by air. Aside from a flock of extinct Carolina parakeets, nothing arriving at my bird feeder could have made me happier. We were discussing the genetics of desert tortoises, something Trip knows far more about than I do, when a bird hit the window. This was only three or four minutes after Trip's latest disparaging remark and chuckle about my "fascinating" bird feeder, which sported only a lone sparrow nibbling away at the free seed. When the sharp-shinned hawk crashed into my plate glass window with the little brown bird in its talons, the reputation of my bird feeder soared.

We both looked out the window with mouths agape as the hawk picked itself up from the ground, regrabbed the twitching sparrow, and gracefully flew away to the pine forest at the edge of the lawn. "That's why I wanted the sparrows," I said, trying to speak calmly. "I knew something like this would happen." Trip was too much a gentleman to call me the liar we both knew I was.

Throughout the fall bird-feeding season, we saw more than half a dozen sharp-shinned hawks rocket to the feeder and depart with a freshly taloned sparrow. We can only wonder how many hawk meals we missed seeing. It was not until the next year that I found out what the victims were doing wrong—eating alone. Anyone with a bird feeder is aware that sparrows often feed in groups. Behaviorists view this as a survival tactic: Approaching predators, such as hawks, are more likely to be seen if several pairs of eyes are scanning the skies. Thus, one bird flying away in fright alerts the entire flock, which can scatter to avoid the danger. Eating by oneself can be costly if one is a fat sparrow in sharp-shinned hawk territory.

All of us know the value of cooperating with other people so that they achieve something they want and we get something in return. Animals other than humans also help each other out, although the effort, as with the sparrows, may not be a conscious one. Many obvious examples come to mind—honeybees that defend the hive, wolves that hunt together in a pack, prairie dogs whose whistles alert others when a predator is near.

Aside from the standard predator-prey relationships, the behavior of individuals or groups is often directed toward members of a completely different species. Sometimes both species profit from a relationship, a situation known to behaviorists as mutualism. A classic example of mutualism is that of the numerous species of cleaner fishes, so called because their food consists of parasites, fungi, and dead tissue from other fishes. Although the small cleaner fishes nibble about on the body surface and even inside the mouth of larger, often predatory species, the latter make no attempt to eat them. The big fishes are rid of pests; the cleaner fishes pick up a meal. Behavioral ecologists attempt to identify the actions of animals and explain them from an ecological standpoint. One objective is to understand why relationships between species initially developed and why they continue to persist.

The relationship known as commensalism is a one-way street. The basic biological assumption in a commensal relationship is that the behavior of one species benefits another, but the first species is not detrimentally affected. A cooperative spirit appears to prevail even if no tangible reward is forthcoming. For example, Spanish moss provides no obvious benefit or harm to oak trees whose limbs support the uninvited guest.

An example of a relationship not necessarily beneficial to both species comes from a study conducted at the University of Georgia. A graduate student, David Young, helped explain earlier observations made by Ronald Pulliam and his associates of an unusual group behavior between rats and sparrows: Rats are more likely to feed in an open area when birds are already feeding there. Cotton rats are fat, brown balls of coarse fur that inhabit the southern tier of states and are considered ideal meals by many predators, including hawks. Thus cotton rats are cautious about venturing into open areas where they might be caught. Nonetheless, if a board covered with seeds is placed in a field, cotton rats will eventually move into the open area to get the free meal—usually on one condition. A flock of feeding sparrows must already be on the board.

In the study, video cameras recorded the response of sparrows and cotton rats to the presence of large plywood squares sprinkled with seeds. Shortly after the food was put out, sparrows began to come and go. And so did the cotton rats. However, the rats would wait along the margins of the board until the sparrows arrived. The rats and sparrows ignored one another while eating, but if the sparrows flew away, the rats would usually scurry off the board.

The investigators concluded that the rats were taking advantage of the sparrows as sentinels and were too wary to venture into the open without benefit of the warning system. The video camera setup confirmed that the cotton rats were aware of the potential danger of feeding in the open. Feeding rate in terms of number of seeds consumed per minute, a measure of success, was determined directly from the videotape. On rare occasions, the investigators videotaped a rat feeding on the board with no sparrows around. The feeding rate of a lone individual was one-tenth that of a cotton rat feeding on the board when sparrows were present. The reason: Solitary cotton rats spent the majority of their time surveying the skies for hawks, leaving little time for dining. Whether sparrows derive any benefit from sharing their dinner table with cotton rats remains to be determined.

An awareness of the biological coordination inherent within ecosystems leads to the realization that large numbers of species interact with each other in a constant and dynamic manner. Interactions among different species in a habitat can determine the makeup of an ecosystem. Biodiversity, and even chemical processes in an ecosystem, may depend on such interactions.

We may as well become accustomed to the term *biodiversity*. Like the word *ecology*, it should, and will, become a part of our everyday vocabulary. Biodiversity, or biological diversity, means "different types of life"; it represents, in essence, the number and variety of plant and animal species or genetic assemblages within any given ecosystem. Biodiversity is an indicator of and contributor to ecosystem health and resiliency. When we destroy or remove plants and animals from a region, even on a local scale, we lose biodiversity. Even if extinction does not occur, the loss is permanent in terms of genetic information. Lowering biodiversity in a habitat not only affects the species removed but also impacts the entire ecosystem by affecting interactions among different species. An important concept along this line, though unfamiliar to many people, is that of *keystone species*, particular species of plants or animals that control the character of an ecological system. Keystone species can dramatically alter an ecosystem's structure and dynamics.

For example, the elephant can create pathways through the jungle, change the quality of a watering hole, and consume enormous amounts of vegetation. Obviously, if elephants are removed from their native area, the entire habitat will change. Removal of largemouth bass can change the character of a lake. With the bass gone, smaller fish will increase in numbers and consume a greater proportion of zooplankton. Zooplankton eat algae, so the amount of algae in the lake may increase noticeably as a result of the absence of bass.

Elephants and bass are obvious determinants of ecological systems, but some keystone species are less apparent. Identifying the importance of some species requires ecological experiments. A study conducted in a desert habitat has shown that vegetation can be dramatically affected by something as seemingly inconsequential as kangaroo rats. A twelve-year study conducted by James H. Brown and Edward J. Heske of the University of New Mexico has shown that removal of kangaroo rats changed a shrub habitat in the Chihuahuan Desert to one of tall grasses. The study was conducted in a transition zone from desert scrub, where shrubs are prevalent, to a desert grassland. The change from shrub land to grassland is dramatic.

Kangaroo rats are little tan rodents with white undersides and powerful back legs. They hop like kangaroos. Like many other rodents, they eat mostly seeds and other vegetation. They dig burrows beneath the soil where they spend the intolerably hot days of summer, and come out at night to feed. Three species live in the area where the study was carried out. The experiment was performed inside two dozen square enclosures of wire mesh, each about 150 feet on a side. All three species of kangaroo rats were removed from eight of the enclosed areas; some plots were left untouched as controls; different types of manipulations were carried out in others. The plots without kangaroo rats proved to be the most affected.

A change in the vegetation was obvious within five years, and after a decade with no kangaroo rats, the habitat had changed appreciably. The amount of tall grasses in the plots without kangaroo rats tripled, compared to the natural vegetation in the plots where kangaroo rats still lived. An entire plant community was altered by the removal of a species that would not even be apparent to a casual observer.

The change in vegetation, as determined by the investigators, was a consequence of the burrowing and food habits of the kangaroo rats. The rodents selectively eat large seeds. Furthermore, they make runways and move vast amounts of desert soil, creating a favorable habitat for many short grass species. Once the kangaroo rats were removed, large-seeded plant species increased and dominated the plots. But the most significant habitat change resulted from the absence of soil disturbance by kangaroo rats. Under the new conditions, the tall grasses had an advantage and became more prevalent.

Other changes also occurred. Six species of small rodents filled the void left by their kangaroo rat competitors. On the other hand, fewer seed-eating birds used the plots with no kangaroo rats, because the increased grass cover made ground foraging more difficult. Also, the investigators noted that on plots without rats, the winter snow persisted much longer, presumably because of the reduced amount of bare ground.

Small, not particularly obvious species can have a major influence on the natural vegetation of an area. Kangaroo rats are not special. Every natural habitat has its influential keystone species, whether we are aware of their importance or not. Even humans are not exempt from mutualistic associations with native species. A research project conducted in Kenya emphasizes this, providing scientific documentation of a fascinating ecological—and sociological—phenomenon. Two scientists, H. A. Isack and Heinz-Ulrich Reyer, investigated and confirmed what many thought was merely an African legend: a bird that leads native hunters to honeybee hives. Although reported in written accounts almost four hundred years ago, the supposed interaction was considered by many scientists to be a myth.

The study provides insight into a strange and ancient relationship between birds and men. According to twenty-thousand-year-old rock paintings from the Sahara Desert, Zimbabwe, and South Africa, humans have gathered honey for at least that long. A mutually beneficial association between humans and an African bird known as the honey guide developed long ago, very likely because both could benefit from and contribute to a raid on a bee colony. The story was that the bird would perform a series of behavioral acts to lead hunting parties to an active bee colony. The men would benefit by being led to a source of honey they probably would not have found otherwise. They obtained the honey by destroying the hive, and the honey guide would get its reward of honeycomb wax and grubs.

To test the presumed myth, the investigators studied a nomadic African tribe, the Boran people. Dr. Isack, a tribe member who did graduate studies at Oxford University, interviewed professional honey gatherers. They told him that Boran hunters summon a honey guide by blowing through their hands or a snail shell. The whistle can be heard more than half a mile away. Soon a honey guide appears, flies close to the whistler, and then hops around on nearby branches. The honey guide begins to call, making sure it has the hunters' attention. Then the bird flies away over the treetops, returns to ensure the men are following, and then leaves again. By following the bird, the hunters are eventually led to an active bee colony.

In the interviews, the Boran honey gatherers claimed that the birds disclose the direction of the nest, the distance to it, and ultimately its actual location. The story of a bird leading humans to a mutual food source was difficult enough to believe. To accept that the bird can give even more detailed information was even harder. So the scientists accompanied the Boran hunters on several searches for honeybee hives. The tales, they discovered, were true.

Establishing that the bird showed the hunters the direction of the hive was relatively easy. The bird's flight path when it left the hunting party was straight toward the nest, which might be more than a mile away. By leaving some honeybee nests intact, the researchers were able to get honey guides to lead them to the same one from different directions. The birds always went straight toward the nest.

The Boran hunters also claimed they could tell how far away the nest was by the bird's behavior. Measurements by the scientists confirmed the claim. The length of time the bird disappears the first time indicates whether the nest is near or far. Later, during the guiding, the distance between successive perches decreases as the hunters get closer to the nest. Finally, when they arrive at the bee colony, which may be carefully concealed, the honey guide perches alongside the nest and gives a special call. As the hunters draw near, the bird flies around the bee's nest to indicate its location.

The story is unusual because it demonstrates a complex ecological relationship between animals and humans. Perhaps we are not far removed as an interactive part of natural environments and ecological processes. A sour note is that this study may have been one of the last opportunities to confirm the honey guide story. According to the investigators, the use of wild honey in much of Africa has been replaced by commercial honey, sugar, drugs, and alcohol. The honey guide birds are still ready to guide, but no one pays attention.

The special attributes and distinctiveness that different species have evolved are fascinating, and those species whose natural histories are inextricably entwined with another's may be the most intriguing of all. Of the thousands of plant and animal species in a region, we seldom know much about the intricate but important relationships that exist between particular ones.

Some organisms depend on others for not only their survival but also their propagation. Research by Eric Dinerstein and Chris M. Wemmer from the National Zoological Park in Virginia documented this fact in Nepal as it relates to the fruits of *Trewia* trees and a species of rhinoceros. *Trewia* trees produce yellowish green to brown fruit about the size of a small apple. The outer coat has the texture of a green potato; the taste is bitter. Birds and monkeys do not eat them, but rhinoceroses readily devour them when the fruits fall to the ground. A large bitter-tasting fruit seems befitting for the personality of a rhinoceros.

The objective of the study was to determine the importance of greater one-horned Asian rhinoceroses in dispersing *Trewia* seeds on floodplains in the region. The general goal was to test an earlier theory that large, plant-eating mammals are instrumental in the germination and distribution of certain seed plants. For example, the African elephant on the Ivory Coast has been credited as the only means of dispersal for thirty plant species. Dispersal is achieved when an animal eats the fruits and then moves to another area before the seeds have moved through the gut and been deposited on the ground.

Some ecologists theorize that the recent reduction or extinction of many large mammals has had a major impact on the dispersal and distribution of a wide variety of plants. Imagine if squirrels and other rodents did not carry acorns away to bury them. New oak trees would grow primarily beneath the parent tree, and the spread of a species would be greatly limited.

Ecologists who study the feeding behavior of rhinoceroses do so with binoculars from the safety of an elephant's back. Rhinoceroses differ from sharks, crocodiles, and tigers by eating plants rather than meat. But rhinoceroses are like some sharks, crocodiles, and tigers in one way: They consider humans to be fair targets for attack. A person cannot stand three rhinoceros-lengths away and watch one eat, as if it were a squirrel in a park. However, rhinoceroses know that elephants are bigger than they are. They do not pick fights with people riding on the backs of elephants.

The investigators actually calculated how many *Trewia* fruits rhinoceroses eat by watching them feed during timed sampling periods. A single rhinoceros eats an average of 197 *Trewia* fruits per day. That's about eleven pounds of fruit. More important to the *Trewia* tree, the herd of sixty rhinoceroses under study was estimated to consume, and deposit elsewhere, more than 1.5 million *Trewia* seeds per year!

The investigators compared the abundance of *Trewia* trees in the Royal Chitwan National Forest in southern Nepal (where rhinoceroses are abundant) and the Royal Bardia Reserve in western Nepal (where rhinoceroses have been extinct for two centuries). In Chitwan, almost half the trees in the study area are *Trewia*. In Bardia, *Trewia* trees are rare, presumably because no mammal the size of a rhinoceros is there to distribute them across the landscape. In some parts of India where rhinoceroses no longer occur, *Trewia* trees are common, but in each case other large herbivores, such as Indian bison or domestic cows, occur in large numbers and eat the fruits. The evidence suggests that the success of the *Trewia* tree depends on rhinoceroses and other large, fruit-eating mammals.

Thousands of such interactions between plants and animals exist, and the only factor limiting the known number is their discovery by ecologists. Even fewer complex ecological interactions between species of plants have been documented. This is partially because very few plants have behavior patterns of any sort that can be observed; most relationships are passive and more subtle than animal interactions. A study in the Sonoran Desert, however, has provided evidence for a phenomenon in which one species of plant affects the survival of another. The relationship turns out to be very important in the desert environment, where only a few plant species are able to survive and thrive.

In the Sonoran and Mojave deserts of the American Southwest, a common plant is the agave, a group that includes the century plant. Agave is widely known as a popular garden plant. The persistence of agave in the desert is impressive, considering the temperatures that must be tolerated. A study by Augusto C. Franco and Park S. Nobel of the University of California, Los Angeles, revealed that all the roots of an agave seedling remain within three inches of the surface. They also found that soil surface temperatures in the Sonoran Desert reach as high as 159 degrees Fahrenheit. How does a young agave plant manage to survive its first full day of summer sun in such an environment?

The investigators found that nearly all agave seedlings were growing in association with another plant species known as desert bunchgrass. And most seedlings occurred only in the center of or on the north, and cooler, side of the bunchgrass plant. The shade provided by the bunchgrass protected the seedlings from the desert sun. Thus desert bunchgrass serves as a "nurse plant" (a term used by ecologists) on which agave depends for its survival during early development.

Examination of the soil also revealed that nitrogen, a critical element for growing plants, is significantly higher around the base of a bunchgrass clump than in open soil areas. Therefore, agave plants also benefit from enriched soil. One detrimental aspect for a seedling agave associated with bunchgrass is that the amount of water available to the roots is reduced, compared with the availability of water at other sites on the desert floor. But the value of the shade, essential for early survival, apparently outweighs the water loss that must be endured by a young agave. In essence, without bunchgrass or another suitable nurse plant, there might be no agave in some areas.

The desert is a relatively simple ecosystem in terms of the number of plants and animals and their interactions. Imagine how complex the ecological network is in an oak and hickory forest or a tropical rain forest with its rich fauna and flora.

Mutually beneficial relationships between species often have a fragile existence, easily disrupted by habitat manipulation. Evidence for this is seen in a discovery made by Julie Wallin, a graduate student at the University of Georgia. She needed a stream not severely affected by urban, agricultural, or industrial pollution. Since pristine streams are becoming a rarity, she conducted her study on the Department of Energy's Savannah River Site. The site, protected from public disturbance, is an ideal location to conduct stream experiments.

The bluehead chub is a small fish common to streams of the Piedmont in areas with gravel. The gravel is critical because bluehead chubs construct their nests from small stones. Several males work together, picking up pieces of gravel in their mouths and carrying them to the nest site. Females congregate around the pile of stones, and the males build spawning pits at the upstream edge. A spawning pit is a depression in the stone pile over which the eggs are released by the females. The waiting males fertilize the eggs, which eventually settle in the gravel nest. After spawning has occurred, the males continue to rearrange the furniture, picking up stones and moving them around in the nest. This prevents the nest from accumulating silt and helps aerate the developing embryos.

A fish using its mouth to build a nest out of rocks is interesting but not too unusual considering the nest-building behaviors of many other animals. What *is* unusual is the association with yellowfin shiners, stream fishes of the Carolinas and Georgia that occur only in streams where nest-building chubs live. During the chub nest-building activities, these smaller minnows begin to gather in the area until hundreds of them form an enormous, frenzied school over the nest. The male chubs for the most part go on about their business, paying little attention to their uninvited audience.

Yellowfin shiners do not congregate around the gravel nest of the chubs only because they like to watch other fish work. Shiners also lay their eggs in the newly built nest, which becomes a safe harbor for the shiner eggs, too. Indeed, yellowfin shiners lay their eggs only in the gravel nests constructed by chubs. The shiners cannot reproduce unless the chubs build a nest for them.

Yellowfin shiners clearly benefit from the relationship with chubs. But do the chubs gain anything from the relationship? The study showed that the shiners potentially contribute to the survival of chubs in two ways. One is through creating a confusion effect around and above the nest. Predators such as other fish and snakes, and maybe even a kingfisher, might be more likely to catch a shiner than a chub. In addition, when shiner eggs are mixed in with those of chubs, it lowers the chance of something eating a chub egg. The advantage to the chubs is subtle but apparently sufficient for them to tolerate the presence of the shiners at their nesting site.

An environmentally educated society must be sensitive to the delicate equilibria necessary within and among species. Species and their interrelationships are products of a long and arduous evolutionary process. When human intervention is too abrupt or too extensive, some species cannot cope in a methodical and programmed evolutionary manner; that is, they have no time to make adjustments by natural selection.

Cotton rats and sparrows, honey guides and honey hunters, chubs and shiners —examples of beneficial and dependent relationships between species could be extended indefinitely. Even a seemingly simple biological phenomenon, such as a rhinoceros eating fruit, may represent a deep-rooted and fine-tuned ecological relationship.

PART TWO

Endangered and Threatened Species: The Specter of Extinction

Natural forces have eliminated local populations of plants and animals and caused the extinction of many species over geologic time. Today's problems stem from the additive effect of human destruction, both of natural habitats and of targeted species. Humans are distinct from all other species: We are capable of destroying members of any taxonomic group (including our own) on a global scale. Even some species that appeared to be environmentally robust were not immune to human environmental attitudes and actions.

With the complete removal of a living species from the earth, only a memory remains, maybe in our minds, maybe in the form of fossils. At best, each extinction event we cause can be considered a training tool, a lesson in how to prevent it from happening again. The reality of the endangered or threatened condition of many modern species is accepted by most environment-minded people today. Most people acknowledge the specter of human-caused extinction, although opinions vary about the gravity of the situation.

A difference exists between today's species extinction threat and the threat of sixty-five million years ago. Then, a natural phenomenon, perhaps the collision of a giant meteoroid or asteroid with the earth, ultimately terminated the existence of countless species, including the dinosaurs. Today we are dealing with humans, a species that can make plans and change the way the world operates. Even if today's biodiversity is significantly reduced, it will eventually return to today's levels through the inexorable course of evolution. This has happened following other periods of mass extinction, but the cast of species is always different from the original. Some of us would rest more easily knowing that our species did not create the need for recovery and that it remained among those still in existence.

Species have fallen prey to the human hand for several reasons: short-term exploitation, as with many of the birds, by those who lacked awareness or appreciation of the consequences of their actions; focused attacks against large predators, such as mountain lions and wolves, because they are perceived as direct competitors or potentially dangerous; and persistent discrimination against certain groups, such as snakes, in response to human fear and ignorance. To change the projected escalation of extinctions in the future, our only hope is a worldwide change in attitude, founded on the knowledge that we can, and should, live in concert with other species, not in conflict with them.

Nowadays, Extinction Is
Usually an Unnatural Act

More than 99 percent of the species that have ever lived are now extinct, mostly because of natural forces. Throughout the history of life on earth, natural forces such as high winds, floods, volcanoes, fires, and drought have reigned supreme in shaping the patterns of colonization, distribution, and inhabitation by plant and animal species. Evolutionary processes made adjustments in the species inhabiting a time and place. However, such adjustments are never truly final and must continue to be dynamic, responding to the vagaries of nature. Environments are subject to natural change, sometimes dramatic, on desolate islands, in serene coastal communities, and even miles beneath the sea. The world's species have their hands full responding to the natural course of events.

Consider what the unharnessed hand of nature can do on a regular basis. Is nature itself an environmental culprit? If you have seen nature's capabilities through the eye of a hurricane or the night glow of molten lava moving down a mountainside, you might think so. The destruction of individuals, the elimination of populations and, potentially, the extinction of species can occur naturally, even today.

Volcanoes can not only eliminate populations of many species and cause the extinction of some, but they can result in environmental changes affecting the entire world. Even the human influences on global warming can be masked temporarily by the eruption of a major volcano. Volcanic activity at the earth's surface has been around since the crust cooled, and it will continue for at least a few more billion years. Everyone in the United States remembers the eruption of Mount Saint Helens and knows that Kilauea in Hawaii still has active lava flows. Many volcanoes of the past also made lasting impressions. Anyone who had a good course in Latin remembers Vesuvius and all the mummified bodies in the town of Pompeii. The King Kong of volcanoes was Krakatoa, about a hundred years ago. In recent times, several volcanoes have received attention. One is Unzen in Japan; another is Mount Pinatubo in the Philippines.

According to *Science* magazine, Mount Pinatubo's eruption in 1991 could possibly be the largest of the century and is of particular interest to scientists. The volcanic eruption resulted in the loss of lives, the emigration of local inhabitants to other regions, and the abandonment of a United States military base. A one-time incident that kills or displaces a few thousand people would be judged as disastrous. However, changing the earth's temperature by half a degree could have a significantly greater long-term effect on the human race. Global cooling from a volcano can occur as a result of sun-shielding by airborne products from the eruptions. The ash, most of which settles to earth within a few weeks, is not the volcanic product that blocks the sun's rays for a long period. The shielding is from tiny droplets of sulfuric acid that can remain in the stratosphere for up to three years.

Whether this will prove to be true for Mount Pinatubo, a volcanic eruption could unquestionably have significant and long-lasting effects on the atmosphere and climates of the world. El Chicon, the massive volcano that erupted in Mexico in 1982, produced enough sulfur dioxide, which combines with water in the atmosphere to form sulfuric acid, to lower the temperature of the earth's surface for a couple of years. This does not mean that either Atlanta or New Orleans was noticeably less hot in August. The total cooling effect was less than a degree, but this is an average for the entire world, a significant influence. Mount Pinatubo is calculated to have spewed out more than twenty million tons of sulfur dioxide that reached the stratosphere, nearly three times that of El Chicon. The projected global cooling effect of Mount Pinatubo could confound studies to determine if global warming from human activities is a phenomenon to be apprehensive about.

A concern of some scientists is that the sulfur droplets from Mount Pinatubo could cause man-made chlorine compounds to deplete atmospheric ozone at a faster rate. El Chicon is estimated to have reduced the amount of ozone in the stratosphere by as much as 15 percent for a short while at certain latitudes. The phenomenon is not totally predictable, but Mount Pinatubo could have a major, though temporary, influence on ozone depletion.

Krakatoa, one of the greatest volcanic eruptions in recorded history, truly had a major influence on the environment. If something of that magnitude occurred today, most political, social, and economic news stories would seem trivial. When Krakatoa blew its lid, an eighteen-square-mile volcanic island located between Australia and Borneo disappeared. Presumably, so did some species. An island of that size and separated by many miles of ocean from the closest land probably had a few endemic species. In the Galápagos, smaller islands than Krakatoa harbor species that occur nowhere else.

The sound of the Krakatoa eruption was reportedly heard more than 3,000 miles away. Tidal waves were created in southeast Asia, and at least thirty-six thousand people perished in coastal cities. Rocks and ash have been calculated to have been thrown more than 15 miles high. The ash cloud was so thick that villages more than 150 miles away were in total darkness for days.

With evidence like this, some might say that nature does more harm than man. Why should we be concerned about causing the loss of habitats and biodiversity, since nature has taken a far greater toll over the eons than humans have? Why be concerned about clear-cutting an old forest, since natural forest fires and hurricanes do the same thing or worse? Why worry about draining wetlands, since natural droughts have the same effect? The answer to these questions is that natural, long-term, persistent environmental stress to a region can be handled by the plants and animals, at least evolutionarily. Most of today's species are unable to tolerate the new and unnatural force, human interference, in *addition* to the relentless forces of natural phenomena. Human destruction and devastation inexorably tip the scales against natural habitats and the organisms that inhabit them.

Clearly, the natural phenomenon of an asteroid strike or a volcano could make most human environmental impacts appear slight by comparison. But this is no excuse for us to proceed with some of our own destructions, aided by a technological edge that produces chain saws, explosives, and pollution in the forms of herbicides, pesticides, and chemical by-products. When we cause massive disruption of wetlands, forests, and marine habitats, the extinctions that ultimately occur qualify as unnatural acts. Scientists argue about the regularity of former, natural worldwide extinctions, and they contest the causes of particular ones. But whatever the schedule or the causes of extinction in the past, today's rate of loss is greater and more rapid than ever before as a result of human activities.

Two things need to be considered. First, if previous mass extinctions did occur, they were caused directly or indirectly by celestial bodies colliding with the earth, by natural phenomena inherent to the planet, or by some combination of such events. Even if human civilization had been in place, humans could not have prevented the destruction. In contrast, today's losses are caused mostly by humans, a presumably controllable force. Second, even the great elimination of species that occurred at the end of the last two geologic eras resulted primarily in the loss of animal species. Plants, the basis of food chains, were not affected so severely. Today, in the tropics, we are rapidly losing plant species as well as animals. Also, even the most catastrophic extinctions took place over hundreds or thousands of years. At our present extinction rate, such losses will be seen over decades. The prediction that we may lose as many as three thousand species of animals and plants per month by the end of the century should give us reason to worry.

Extinction is not always abrupt and dramatic. The processes leading to extinction can occur gradually and naturally, without human influence. Normal ecological succession of a habitat from one type of plant community to another brings environmental advantage to some species and disadvantage to others. Habitat alteration resulting from natural forces destroys individuals of some species and lowers the resistance of some populations to environmental pressures that might otherwise have been inconsequential. Widespread habitat change resulting from shifts in climate, as well as more-localized wind, fire, flooding, and drought, can weaken the chances of survival for some species. Extinction is an unlikely consequence of any one incident, but the cumulative and collective impacts of these natural phenomena can influence species composition and even result in the extirpation of some species in a region.

People generally show great concern over powerful winds and flooding because of economic and human interests, not because of the severe effects on natural systems. Most news coverage of natural phenomena such as hurricanes or volcanoes concentrates on property damage and the human consequences. Because of this focus, the environmental side of such events may seem trivial or be underemphasized.

Hurricane Hugo, the autumn storm that destroyed human habitation from the Greater Antilles to the Atlantic Coast in September of 1989, took a mighty toll on certain wildlife. The environmental lessons were many. In one sense Hurricane Hugo taught a full course in storm ecology. High winds and flooding caused by hurricanes can quickly reduce the population sizes of some species in a region. This is to be expected. Animals and plants native to hurricane regions have died from winds, rains, and flooding for centuries. The real lesson is that representatives of every wildlife species typically manage to survive, if for no other reason than because of their abundance elsewhere.

A few weeks after Hurricane Hugo, the U.S. Fish and Wildlife Service released information on the storm's effect on certain wildlife species. For example, eighteen out of nineteen bald eagle nest trees along the South Carolina coast were destroyed. Although the bald eagle is in a less precarious position than some endangered species because of its wider geographic distribution, Hurricane Hugo dealt a major blow to this majestic species in the region. Another bird species was more severely affected, to the point of impending extinction. The wild parrot population of Puerto Rico was estimated to have been virtually eliminated. Only six of about fifty parrots living in natural habitats being studied were located after the storm. The species was assumed to be on the brink of extinction in the wild. Fortunately, two years after Hugo, the wild population showed substantial recovery. But it was clearly a close call.

Why would a single storm threaten the existence of a species like the Puerto Rican parrot? Partly because the species is now confined to a small geographic range and has been reduced to small population sizes because of earlier habitat destruction by humans. When it becomes apparent that a species might be eliminated entirely by the natural disaster of a hurricane, we should take it as a warning signal: The species may be struggling to endure the impacts of *both* humans and hurricanes.

The red-cockaded woodpecker, also endangered like the Puerto Rican parrot and bald eagle, represents another species severely affected by Hurricane Hugo because humans had already made it susceptible to extinction. Red-cockaded woodpeckers have highly specific habitat requirements. They nest exclusively in old, living pine trees, around which they center their colonies; the favored trees are at least eighty years old. The woodpecker chisels a cavity in the tree, usually twenty to fifty feet above the ground. Most cavity trees are also infected with a fungus that creates a condition known as red heart disease. The fungus softens the pine heartwood and may be the critical factor that allows a red-cockaded woodpecker to peck a hole into an otherwise very hard tree. In the cavity, the woodpeckers raise their young and spend the night and other inactive periods. Imagine trying to endure a hurricane or tornado if you lived in a tree house!

These nonmigratory woodpeckers live in small family groups called clans. A clan usually consists of a mated pair, bachelor male offspring, and several juveniles. The young males assist in defending the territory, incubating eggs, and feeding nestlings. The nesting cavity is critical to the natural history of a red-cockaded woodpecker clan. Because of the special requirement for aged, disease-weakened pine trees with the right qualifications, cavities are at a premium. A cavity is often passed down from one generation to the next, and each clan's territory usually consists of several cavities.

Red-cockaded woodpeckers took it on the chin from Hurricane Hugo. The year before, 466 clans were known to live in the Francis Marion National Forest, pine savanna woodlands along the South Carolina coast north of Charleston. Following the storm, only three colonies of clans remained completely intact. As many as 75 percent of the birds were estimated to have been lost. But wait. If 466 clans had existed in the first place, one might ask why the species, which is found from North Carolina to Texas, was ever considered to be in trouble.

Peter Stangel of the University of Georgia's Savannah River Ecology Laboratory studied the birds for several years. According to him, the red-cockaded woodpecker is considered endangered not only because few birds remain, but because those left are found in isolated and often small populations. Except for the Francis Marion population, most of the remaining two thousand to three thousand red-cockaded families are found in colonies of fewer than twenty clans each. Such small, tightly clustered populations are vulnerable to focused environmental catastrophes like hurricanes and tornadoes, as well as to certain forestry practices, that can destroy old trees.

The woodpecker losses in the large population of the Francis Marion forest are striking for several reasons. First, they illustrate that even very large populations cannot always endure natural disasters. Second, the population represented 15 percent of all red-cockaded woodpeckers in existence. And third, this was the only large population that had increased in number of families during the last eight years. In all other areas censused, the birds had either declined in number or remained stable. Unfortunately, forestry practices that remove the oldest trees in a pine forest seriously threaten many of these habitats. Red-cockaded woodpeckers in such habitats are destined for local extinction.

The Francis Marion population of red-cockaded woodpeckers was so healthy before Hurricane Hugo that scientists had removed birds to transplant elsewhere. Eight were taken to the U.S. Department of Energy's Savannah River Site near Aiken, South Carolina, in an attempt to build up the small but protected population of woodpeckers there. The two dozen or so now living on the site appear to be doing fine, but the fate of the species in one of its last major strongholds, the Francis Marion forest, is now in question.

Native wildlife species will very likely survive hurricanes as long as they do not also have to deal with problems created by us. Morally, we need not have concern for the welfare of species living in their native habitats and dealing with environmental extremes from natural causes. This is the way of the wild. The problem today is that the wild is for the most part gone, and most existing species are affected in some way by human activities. Their guard has been lowered in fending off the assaults of technology and human intervention.

Forest fires are other natural events classed as disasters by humans, especially if human lives are lost or economic loss occurs in the form of destroyed property or merchantable timber. But in regions where fires are to be expected from natural events such as lightning, nature is prepared. Many temperate-zone habitats are classed as "fire climax" systems, which means that the dominant vegetation is composed of species that are fire resistant or fire dependent in one way or another. For example, in the upper midwestern United States, fire causes the cones of jack pines to open and the seeds to disperse. The accompanying burning of the litter layer exposes mineral soil and optimizes seed establishment.

Longleaf pine in the southeastern United States takes several years to pass through the so-called grass stage. During this stage a future majestic pine tree resembles a tuft of bright green grass but has its major reserve in a below-ground root system. Thus a young longleaf pine can withstand the periodic lightning fires that occur naturally. Although the outer needles may burn away, the major portion of the plant remains safe below ground, ready for regrowth in the following season. A side benefit of fire is the return of essential nutrients to the soil, making them available for use by plants.

If vegetation in a region is fire tolerant, rest assured that the indigenous animal species will have also adapted to living with periodic fires. Mortality in most vertebrate populations during natural fires is generally considered to be negligible, although small birds and a variety of mammals and reptiles have been reported killed in forest and prairie fires. Invertebrates may suffer immediate losses of eggs and larvae, but recovery to normal population sizes generally occurs for most species over a few months.

56   Under prehuman natural conditions, fires roared through forests at frequent enough intervals to prevent buildup of the ground litter. Without fire fighters to extinguish them or roads to retard their spread, some natural forest and prairie fires probably covered immense areas. Natural fires were extinguished by rain or when a river or other aquatic habitat served as a barrier. After a natural grass or forest fire, little or no flammable material remained. Only a scant covering of dead grass, leaves, or pine needles would be available to serve as fuel in the years following a fire. Thus future fires would burn rapidly and usually low to the ground. Fire resistance was enhanced by the thick bark of big trees whose upper branches were out of reach of a low, fast fire. The root reserves of low-growing plants assured their quick recovery. All served to maintain a healthy forest. Once the cycle was established, fires were simply part of the natural environment, no less expected or more damaging than an occasional cold winter or major flood.

However, in areas where fire has been controlled for decades, the litter buildup can be immense, perfect fuel for a hot and prolonged fire that is more likely to reach the tops of tall trees and develop into a crown fire. In addition, the heat penetrates deeper, raising ground temperatures and burning not only the soil but also the plants and animals living below the surface. Many of today's hazardous forest fires occur in habitats that have been unnaturally protected from fire. When one does occur, it can be devastating to both wildlife and humans.

Entire generations of schoolchildren have been brought up with the notion that forest fires are bad. Indeed, a fire started from an ill-tended campfire or discarded cigarette is unnatural and can be catastrophic. But, except for certain instances of human influence, forest fires are as natural as hurricanes and tornadoes. Even the impressive fire that burned 1.4 million acres in and around Yellowstone National Park in 1988 is considered by ecologists to have been a normal event. Tree records reveal that forest fires of the same scale and intensity occurred in the region in the 1700s, before the influence of European settlers.

Only humans perceive fire as an environmental disaster. Major efforts are made to control fires because of timber and domestic interests, not to protect native wildlife. Ironically, unnatural fire control has been the major detrimental human impact on natural systems whose flora and fauna have evolved to deal with periodic fires. Management policies that fail to recognize the role of occasional fire in sustaining fire-adapted species may increase the risk to humans as well as to native wildlife. In regions of the world where fires occur naturally, the wildlife and plants have acquired mechanisms to persist.

Drought is a natural stress that can be long-term and geographically widespread. The universal requirement for water by plants and animals, including humans, results in some of the most dramatic longer-term natural disasters. Drought and subsequent famine have been recorded since biblical times, and continue today. We are besieged by news broadcasts on the plight of regions or entire countries undergoing the consequences of drought. As with other natural and man-made disasters, the newsworthy items generally relate to conditions being endured by the local human population.

Native plants and animals also undergo stress during a drought. Some animals migrate to more favorable areas if they exist in the region. Some amphibians and reptiles burrow beneath the soil and remain dormant for months, perhaps years, if necessary. Trees and other plants may die, but most produce seeds that can wait indefinitely until suitable moisture conditions return. If drought is a naturally recurring event in a region, the indigenous species have the means to survive. Wildlife species that inhabit the ultimate drought ecosystem, the desert, persist under what can be regarded by humans as stressful conditions. But to the organisms that live in a desert, drought is normal.

Many modern drought situations defined as problems are those in which the natural systems and native wildlife of a region could have persisted had human activities and demands not exacerbated the problem. The impact of the human quest for water can sometimes weaken the natural resilience of an ecosystem and push it to the breaking point. For example, bringing water from up to five hundred miles away to a semiarid region such as southern California is a short-term solution. The excessive removal of water from natural environments seems an unacceptable solution to human overpopulation in a region, especially when the perceived "needs" include water for golf courses, lush lawns, and clean cars. Developers must face the reality that unlimited growth and development is not a healthy option in places where there was not enough water to begin with.

Natural disasters are usually temporary, their detrimental effects eventually ceasing; and they seldom occur with an intensity that brings worldwide devastation. In contrast, human impacts are not on the wane, and the reverberations are being felt throughout the world. Early humans, using spears and knives, probably wrought extinction for a few species. But at that time, as a consequence of natural events such as volcanoes and ice ages, extinctions probably occurred throughout the world at a more rapid pace than those caused by the nomadic bands of a sparsely scattered, two-legged species carrying small weapons. Today that same species occupies all continents, and its arsenal includes bulldozers, dredges, explosives, and chemicals. The foremost weapon is a knowledge of how to use these tools to exploit any natural resource on earth.

We have altered the planet to an unrecoverable point from the ancient era when we first achieved our special dominance over other life. Only recently, however, have we passed Nature herself in the rate at which species are added to the roll call of the extinct. The interactive effect of human destruction and natural environmental extremes has become overpowering for more and more species. Natural forces are not going to stop operating. Plants and animals will have to continue contending with hurricanes and fires, volcanoes, droughts, floods, and more meteoroids. Let's not force them to have to cope with us, too.

Birds and the Ecovoid

I think we need a new word in our environmental vocabulary: *ecovoid*. I define an ecovoid as a missing component of our environment that we wish were present but can never be replaced. Consider, for example, blue jays—loud, sassy, colorful. Most people take them for granted, but if they were to disappear, many of us would feel that a severe ecovoid had been created. If I knew I would never see, never hear, another blue jay, I would feel that something was missing from my own existence.

Two ecovoids were created in the early 1900s. The first occurred in 1914 when a bird named Martha died in the Cincinnati Zoo. Martha was the last known surviving passenger pigeon. When she died, her species became officially extinct. The term *rare and endangered species* suggests that plants or animals on the way to extinction were never very common. A minor loss, some might say. But passenger pigeons are claimed by some to have occurred in greater numbers than any other bird or mammal for which we have records. For the passenger pigeon, safety from extinction did not come in numbers.

Passenger pigeons resembled mourning doves, a close relative that is still common today. One distinction, however, was that passenger pigeons depended on large numbers for communal nesting. This trait made them easy prey for human hunters and ultimately led to the demise of the species. The abundance of passenger pigeons was documented by both scientists and laymen. John James Audubon, acclaimed for his accuracy in recording natural phenomena, reported an enormous flock of passenger pigeons in Kentucky. According to his records, the migrating flock was more than a mile wide, closely compact, and passed overhead during the daylight hours for three full days. He estimated that more than one billion birds were in the flock.

The abundance of passenger pigeons was further noted in one of the communal nesting areas near Petoskey, Michigan. Almost every limb on every tree in extensive areas of forest had at least one nest. According to early accounts researched by I. Lehr Brisbin of the Savannah River Ecology Laboratory, temporary campsites were set up each year near this nesting colony. Hundreds of people exploited the Michigan passenger pigeon colony, the largest one known. In 1878 the colony was twenty-eight miles long and four miles wide. Countless pigeons were sold, dead or alive, during the late 1800s. Most were used for food. But the birds were extremely docile and cooperative in captivity, and more than twenty thousand of them were used as shooting gallery targets on the Coney Island midway. Despite the presence of millions of birds, the population size dwindled away over the years as the onslaught continued, particularly from commercial "pigeoners."

One way to capture pigeons was to lure them to a site with a decoy. Pigeoners would set a net trap alongside a feeding area where flocks of passenger pigeons were known to pass in flight. A tame pigeon, or stool bird, was used to catch the attention of the passing flock. Upon seeing the decoy pigeon on its stool-like perch, the flock would divert its flight. As the birds landed in the would-be feeding spot around the stool pigeon, the pigeoners immediately netted the entire flock. According to one authority, approximately ten nettings of about twelve hundred birds each were made in a day. In a good week, pigeoners could capture more than eighty thousand passenger pigeons. The actual toll was even greater than this. Because most of the trapping was done during the nesting season, innumerable nestlings lost their parents and starved in the nest.

The extinction of the passenger pigeon is a commentary on a human attitude that continues to persist: the dangerous belief that we can and should exploit any natural resource to the fullest without regard for the future not only of the particular plant, animal, or environment but of our own descendants as well. Selfishly squeezing everything we can out of natural areas for quick financial gain may be the most costly feature of free enterprise. And the final payment may be more exorbitant than anyone anticipates.

60

Some people in the nineteenth century, including a few legislators, recognized the danger inherent in this attitude. By the early 1900s laws were being passed to prevent the wholesale killing and trapping of passenger pigeons. But, as with many of today's environmental laws, the rulings were passed too late, were not stringently enforced, and left too many loopholes. Thus Martha, the last lonely passenger pigeon, passed away in captivity on 1 September 1914.

At that time, only a few feet from Martha's cage, another North American species of bird was showing the last sparks of survival. This bird, named Incas, was the last confirmed specimen of the Carolina parakeet. The next avian ecovoid came four years after Martha's demise. On 21 February 1918 Incas died.

The bright green Carolina parakeet was unique among members of the parrot family; it was the only truly North American species. The last sightings in the wild were made in Florida and South Carolina, but the colorful bird with its orange and yellow head had once been reported throughout much of the eastern United States. Though it was predominately a southern species, northern sightings were not uncommon in the early 1800s. Few records were kept of the habits and ecology of Carolina parakeets, but most accounts agree they traveled in flocks, ate fruit and seed, and were noisy. Their natural history is now a matter only for speculation based on our knowledge of how other parakeets behave today. We will never know for sure.

Even the time and cause of extinction of the species is uncertain. Hindsight leads to a belief that their flocking behavior, conspicuousness, and approachability made them easy targets for men with guns. Pretty green feathers for ladies' hats prompted hunters to kill them for sales to merchants, and the birds' propensity for feeding on farm crops caused them to become targets of farmers. Even the scientific community had a hand in their demise. In 1904 a museum curator in Florida shot down four individuals of the last confirmed flock. He watched the remaining nine fly away toward Lake Okeechobee. We can only conjecture whether humans actually changed something about the birds' environment that resulted in their disappearance. But the suspicion is that the Carolina parakeet became extinct because people took them for granted and were either unaware of or uninterested in their gradual decline. Suddenly, it was too late.

But in 1935 exciting news came to ornithologists and bird lovers, perhaps the only ones who really cared at the time. The Carolina parakeet, presumed extinct,

61

had been given a reprieve. A flock had been sighted in the Santee River swamp of South Carolina! The discoverer's credibility was enhanced by his report at the same time of another rare species, the ivory-billed woodpecker, a sighting later verified by professional ornithologists. But despite additional claims by other amateur bird watchers, the confirmation of the Carolina parakeet was never officially accepted. Today we can seek no solace in a possible resurrection of the extinct Carolina parakeet, even if it was in the Santee River swamp in 1935. Almost all of that swamp is gone, as are most swamps in the United States. Even the last four specimens shot by the museum curator in 1904 have vanished. And no one knows what happened to the body of Incas. Only a few stuffed specimens remain, kept in cases or used in displays in museums such as the Smithsonian Institution and the Charleston Museum.

We won't get another chance with passenger pigeons or Carolina parakeets, but maybe their downfall will help instill in us an important awareness: The delicate nature and fragility of some species may be hidden behind a facade of large numbers and apparent well-being. Whatever else we may think about the protection of wildlife and natural environments, we should never take them for granted, especially when we do not thoroughly understand their ecology. Many species may well be on the brink of extinction without our knowledge.

Some people are alive today who actually saw a living Carolina parakeet or passenger pigeon, at least in captivity. Another ill-fated bird is beyond the memory of any living human. The dodo disappeared from earth long ago on the faraway island of Mauritius.

Islands are often ecologically unique. For example, the Galápagos Islands in the Pacific Ocean off South America's coast are noted for their giant tortoises. Ascension Island in the Atlantic is the nesting area for thousands of sea turtles, some of which travel from American coastlines more than two thousand miles away. And everyone knows of zoological oddities such as the kangaroos and duck-billed platypuses of Australia, the world's largest island. Among scientific circles, the island of Mauritius is of particular interest because of the revelation of an often-suspected but seldom-proven ecological relationship. The relationship, known as mutualism, is one in which one species becomes dependent upon another for its existence. The example from Mauritius involves an extinct bird, the dodo, and a nearly extinct tree. Mauritius is located in the Indian Ocean about five hundred miles east of Madagascar, off the southeastern tip of Africa. The uninhabited island was settled by European explorers in the late 1500s. The dodo was only one of the unusual animals on the island. Unfortunately, after settlement of the island by humans, the dodo's extinction soon followed.

Dodos weighed up to thirty pounds (the size of a large turkey) and were completely flightless. Thus they were susceptible to the behavior patterns and ecology of humans. Early explorers and the later settlers could easily catch them for food or simply out of curiosity. Several were displayed in zoos in England and in Europe. To make matters worse for the dodos, they probably had no natural predators, so they made no attempts to protect themselves or their eggs from the newly introduced dogs, cats, pigs, and rats. The last living dodo was recorded in the year 1681.

A recent discovery gives a valuable lesson on how we stand to lose more than the obvious when we exploit the environment without understanding its ecology. The island of Mauritius was absolutely devastated by timber cutters in the 1600s. Among the trees used for lumber was one known as the tambalacoque, a large tree found nowhere else in the world. Several years ago, only a few dozen of these trees, remnants of an ancient stock, were reported to remain on the island. None of them were young trees; all were judged to be more than three hundred years old. Some authorities believe that no seeds of the tambalacoque have germinated naturally within the last two or three centuries, possibly since the last dodo waddled across Mauritius. The reason? The fruit is large and edible, but the seed is covered by a thick shell. Although native animals now living on Mauritius consume the fruit, none eat the seed. But dodos, with their huge beaks, did indeed eat the tambalacoque seed, along with the fruit.

Based on laboratory experiments in which turkeys were fed the seeds, Stanley Temple of the University of Wisconsin concluded that the seeds were able to break dormancy and germinate only after passing through a bird's digestive tract. That is, passage through the gizzard of a dodo may have been essential. The germination of seeds was presumably enhanced by abrasion of the tough outer covering, and seed dispersal was guaranteed. With the discovery of the need to weaken the outer seed coat, small tambalacoque trees are now growing on Mauritius. According to some observers, they may be the first since the extinction of the dodo. For ecologists, one mystery is, why did the tambalacoque tree develop such a dependency on a single animal? How did such a coevolutionary relationship arise? Temple suggested that the fruits had evolved a tough seed covering that could withstand the grinding action of the dodo gizzard. Concomitantly, the plant became dependent on abrasive action for germination.

Not all authorities agree that seed germination in the tambalacoque was contingent solely on the dodo. Scientists, however, often do not reach accord even about the ecology of animals that live today; they rarely do about those that lived before our time. John Iverson of Earlham College has suggested that an identical obligatory mutualistic relationship may also have existed on Mauritius between the tambalacoque seeds and one or more species of now-extinct tortoises. Whatever the true explanation, interactions between plants, animals, and the environment can be far more fragile than we may perceive. The eternal loss of any species should give us cause for reflection.

To be unduly critical of the European explorers who discovered Mauritius and unwittingly destroyed the dodo would be a form of environmental chauvinism. The list of human cultures that could be held in contempt for untimely extinction of other species is probably far greater than we will ever know. Only the largest and more conspicuous species were likely to be noticed and, therefore, to cause comment when their eventual, permanent absence became apparent. Small organisms insignificant to humans might vanish without anyone's noticing their extinction.

But size, like large numbers, is no guarantee of survival. Four hundred years ago, birds larger than ostriches lived in New Zealand. Moas were enormous flightless birds that stood more than ten feet high, but their large size actually became a distinct disadvantage to them. Because moas were big, they were an attractive food item, and because they could not fly, they were unable to escape a new predator, humans, that arrived on New Zealand by the 1300s. Presumably moas became extinct because they and their eggs became preys of the Maoris, the original Polynesian inhabitants of New Zealand.

Moas were big, to be sure, but the elephant birds of Madagascar were larger still. Some individuals weighed more than a ton, and their eggs were the size of a basketball. Exactly when elephant birds became extinct is uncertain, but it was recent enough that nonfossilized eggs have been found in this century. The eggs were of such size and shell thickness that natives cut them in half to make gallon-sized bowls. This was presumably after preparing an omelette big enough to feed a small village.

Again, considering the circumstances and the times, to accuse the Polynesians

64

or the Madagascans of environmental insensitivity would be unfair. A far more disturbing consideration might be that a disregard for the well-being of other species is genetically entrenched across all human populations and cultures. These birds, after all, are just a few examples of creatures that have become extinct as a direct result of human intervention. We have reached a stage in human development when we should absolve the peoples of the past. At the same time, we should set a more deliberate course for the future. The existence of countless species depends on our ability to improve the way we respond to the world. Our own survival depends on it, too. A spark of enlightenment in this regard is seen in the return of the whooping crane.

More than two hundred whooping cranes are alive today, most belonging to a population that migrates twenty-five hundred miles from Texas to Canada. One reason for their existence is a government organization from which we actually get our tax money's worth: the U.S. Fish and Wildlife Service (USFWS). We probably get more than we pay for. Although I can't vouch for every individual in the organization, at most levels, ecologists working for the USFWS are knowledgeable about fish and wildlife, including nongame species and plants. One of the USFWS success stories has been the whooping crane.

Adult whooping cranes are magnificent wading birds that would catch anyone's attention. White, with black wing tips, the whooping crane is the tallest bird in North America. Adults stand up to five feet tall. In flight, with their long necks fully extended, the birds make a long, sonorous sound that can be heard a mile away. Like many other wading birds and waterfowl, whooping cranes spend the winter in aquatic habitats in the South, then migrate north in the spring to nest. At their breeding grounds, they engage in dramatic and elaborate courtship dances and rituals before mating.

Some of the historical facts about whooping cranes are sobering. For instance, migratory whooping cranes once nested as far south as Iowa and Illinois, yet the last known nest in the northern United States was found in 1894. The Migratory Bird Treaty Act of 1916 made hunting the cranes illegal, but uncontrolled urban and agricultural development eliminated all nesting sites in the country. The nesting area of the few remaining birds was not discovered until almost forty years later, and not in the United States.

In 1922 Canada established Wood Buffalo National Park in Alberta and the

Northwest Territories, within a few hundred miles of the Arctic Circle, to protect native wood bison. The reasons they set aside a protected area bigger than Connecticut had nothing to do with preserving whooping cranes. A helicopter pilot, Don Landells, accompanied by G. M. Wilson, accidentally discovered the cranes' last remaining nesting site there in 1954. The wild cranes now breed in Canada and spend the winters in and around Aransas National Wildlife Refuge, Texas, which was established in 1937.

66

Only twenty-one whooping cranes were left in the world when the United States entered World War II. Not surprisingly, the whooping crane was near the top of the list when the first version of the Endangered Species Act was passed in 1966. In 1967 a captive-breeding population, reared from eggs taken from nests in Wood Buffalo National Park, was established at Patuxent Wildlife Center in Maryland. This constituted the only captive flock in existence and reached a population size of about seventy cranes. Thirty members of the flock were moved to the International Crane Foundation in Wisconsin in 1991; now a second major captive-breeding population exists.

In 1975 the USFWS became involved in another project to increase whooping crane numbers. An experimental wild population was started at Grays Lake National Wildlife Refuge in Idaho. Whooping crane eggs were placed in the nests of sandhill cranes, which served as caring, though unsuspecting, foster parents. The foster fledglings, naive about the identity of their real parents, followed the sandhill cranes to New Mexico their first winter and now return there each year. For various reasons, of the more than eighty that fledged during the first fifteen years, only a dozen survived. And not one has yet bred or appears likely to. James C. Lewis of the USFWS indicates that the male whoopers appear to have normal sexual appetites but that the females are attracted to male sandhill cranes, who have no interest in a relationship. No one knows the true biological explanation for what happened. Whatever the case, the project has been suspended. The USFWS will try another approach with the establishment of a new nonmigratory population on the Kissimmee Prairie in Florida in 1992.

The USFWS projects on whooping cranes represent a substantial effort to ensure the survival of a few birds, but we need to show concern like this for a lot more species if we want to maintain an acceptable level of biological diversity. Whooping cranes owe their survival to the concomitant preservation of habitats at

Aransas National Wildlife Refuge in Texas and Wood Buffalo National Park in Canada. Clearly, some species require more than one ecosystem for survival, often ecosystems that are markedly different. For example, many of the birds that inhabit our woods and backyards during spring and summer spend the winter in tropical forests. The tropics represent the savings account of the world's biodiversity bank. Yet almost 40 percent of what was once tropical forest is now gone, primarily due to clear-cutting and other human activities. At the current rate, the last tropical forests will disappear before the middle of the next century.

The destruction of tropical rain forests will have a dramatic impact on the rest of the world, for as these forests disappear, so will our birds. The process is gradual, easy to ignore in the early stages. But imagine the summer day when you realize you haven't seen a hummingbird or a goldfinch all year. And that you may never see another one. Consider how few bird species will come to your grandchildren's backyard the year the last rain forest disappears. We already have enough ecovoids to last a lifetime. Let's not create any more.

Where Have All the
Panthers Gone?

I once wrote a newspaper column about mountain lions, which are also called panthers, pumas, or cougars, and received several letters from not-so-happy people who thought they had seen a large, long-tailed cat. In the column I did not say that no one had seen a wild mountain lion; I only said that, outside of Florida, no recent authenticated records exist for the big cats in natural settings from the Mississippi River to the Atlantic Ocean.

The problem that this statement causes for people who think they have seen a mountain lion lies in the word *authenticated*. Solid evidence is required before the U.S. Fish and Wildlife Service (USFWS) or a state game department will accept a mountain lion sighting as authentic. The evidence must be a photograph, a footprint, or an identifiable body part. Many respondents claimed to have sighted a mountain lion somewhere in the mountains of Tennessee or the swamplands of Georgia and Alabama. Even Pennsylvania, Massachusetts, and Virginia were cited as areas where the big cats still lived—whether I knew it or not! Everyone seemed to think I was questioning his or her credibility.

One report I received was local, from a small town in South Carolina. A seasoned woodsman called to say he had just seen a large, long-tailed cat run into a swampy, wooded area along the Savannah River. He had seen plenty of bobcats and knew he would not mistake one cat for another. He was familiar with the laws about protected species and knew better than to shoot the animal. He also knew that to qualify as a verified record, some tangible evidence had to be produced. Fortunately the animal had left distinct footprints when running away into the muddy edge of a swamp; the evidence was there.

I asked the local U.S. Forest Service personnel if they would examine the footprints. The foresters brought their cast kit, poured plaster into the print, and made a replica. They were able to verify what the hunter had seen: a large dog. Apparently, in his zeal to see what he wanted to see, he mistook a large, long-tailed dog for a mountain lion. Another report I heard about was from a wildlife officer who was called to pick up the carcass of a mountain lion killed in Tennessee. When he and other game officials arrived to get the specimen, they were awed. But not because the specimen was a mountain lion. It was a yellow house cat that weighed thirty pounds—an impressive record in its own right! Had wildlife officials not checked out the specimen, the citizens in that area of Tennessee would still be discussing the mountain lion that had been killed. And it probably would be remembered as a big one. It's a well-documented fact that specimens of many species, including fish and big predators, grow larger after they are dead and gone.

Mountain lions once ranged from Canada to Argentina, including the entire United States. In North America they are still common in some of the wild areas remaining in Canada and the western United States. When the last mountain lion was seen in most regions is not certain. A large individual was killed around Savannah, Georgia, in 1902, and reports of their presence in the Okefenokee Swamp of Georgia in the 1920s are considered valid. Some authorities believe that the frequent sightings still reported from more than half a dozen eastern states are evidence that wild populations do still live in these regions. Others feel that although mountain lions occur in some areas in the East outside of Florida, the animals are escapees from zoos. Most authorities probably do not believe the accounts at all. Among the reasons for game officials' incredulity about sightings of mountain lions are the many cases of misidentification. This is similar to the problem with such phenomena as the Loch Ness monster, Big Foot, and flying saucers. Numerous false reports discredit potentially reliable ones. The most efficient attitude is to assume that none of the reports are true until supported by tangible evidence. That is not always the most popular position to take, however.

Despite the misidentifications by well-meaning people, I like to think that mountain lions still live in the untamed wild lands in the eastern United States, that all reports of sightings are valid, and that the big cats have just gotten better at hiding from us over the last century. I, for one, like the mountain lion and hope it does not represent one more species that has almost been eliminated by our enthusiasm for shopping malls, extensive highway systems, and other habitat modifications we call progress. Obviously, if my feelings were unique, fewer people would be so eager to have seen a mountain lion. The fact that many people want to see these animals where they are not supposed to exist makes a strong statement about ecovoids. Many of us do not like the thought that something our ancestors lived with is no longer around. We do not want to accept it.

Maybe we should be satisfied that mountain lions still persist in some regions and try to protect what we still have. An estimated thirty to fifty mountain lions live in Florida, where they are referred to as Florida panthers. They represent the only well-documented population in the East. According to Sonny Bass, research biologist at the South Florida Research Center, Everglades National Park, this population is located south of Lake Okeechobee, in and adjacent to Big Cypress Swamp. Sightings of individuals have been verified in northeast Florida (Putnam and Flagler counties), but these are presumed to be escapees or lone stragglers from farther south.

Today, because of large-scale habitat destruction, many plants and animals, like the mountain lion, have been eliminated from parts of their geographic range and pushed to the edge of extinction. Their livelihood now depends on marginally suitable habitats that are further threatened by hastily considered construction projects or other environmental alterations. Highways cause particular problems for many animals. Untold numbers of wildlife are accidentally killed annually by motorists, especially at night. Highways also result in habitat fragmentation, which can genetically isolate small populations and lead to inbreeding.

The highway problem is recognized for the Florida panther; at least seventeen deaths from cars were documented from 1972 to 1991. To protect the animals, a nighttime speed limit of forty-five miles per hour is strictly enforced in key panther habitat. A driver who hits a panther while speeding in these areas may find that the cat is not the only victim. Not only can the driver be injured in the accident, he may also face heavy fines for killing an endangered species.

One approach toward protecting animals that tend to cross highways is the construction of wildlife underpasses. The technique has been confirmed to work for panthers. The crossings built under major highways help reduce road kills of panthers, alligators, and other wildlife. Fences are used to funnel the cats and other animals into the crossings so that they can reach the other side safely. Because panthers roam many miles while hunting and searching for mates, providing safe access to as much habitat as possible is important.

The USFWS has developed a Florida panther species survival plan. The plan is intended to prevent this endangered species from going extinct in its last stronghold in the eastern United States. It has elements of a good news-bad news scenario, heavy on the bad news side. The good news is that three or four dozen of the big cats still roam wild in Florida. Furthermore, the USFWS believes that if protection measures are enacted, the panther population size could be increased to 130 by the beginning of the next century and to 500 by the year 2010. The plan involves removing panthers from the wild and raising them in captivity. The bad news is that without a captive-breeding and reestablishment program the USFWS predicts extinction for the Florida panther in less than forty years.

The captive-breeding plan for the cats includes removing offspring of selected adults from the wild annually for three to six years. No more than half a dozen kittens will be taken in a year. Over the next ten years, this breeding stock will be used to increase the numbers in captivity.

The really bad news is that the decline in numbers of mountain lions and the disappearance of other wildlife are being caused by the same thing—human encroachment. At least three major impacts can be identified. First, panthers were once hunted in Florida, an activity that eliminated them from much of the state. Second, their major food source, deer, was hunted. With their primary prey reduced to low numbers, panthers turned to other food sources. But rabbits, raccoons, and possums could not support so large a population of panthers, and their numbers declined further. Finally, habitat destruction took its toll. This has been the most devastating problem of all for the panther as well as for all other native species, especially in Florida. Drained swamps, clear-cut forests, agricultural fields, and highways are not in the best interest of most native plants and animals, especially a wide-ranging predator like a panther. When asphalt and orange groves replace forests and swamps, the natural biodiversity of the area is decreased. Top predators, like panthers, are often the first to disappear.

More bad news for the panther continues from an unexpected quarter. High mercury levels have been found in panthers from a particular region of the Everglades, and mercury poisoning is believed to have contributed to at least one panther death in July 1990. According to the USFWS, the area has fewer deer for the panthers to prey on; thus they more frequently take raccoons, other mammals, and small alligators as food items. The USFWS suggests that these smaller animals, linked to an aquatic food chain, may accumulate mercury in higher concentrations than do deer.

Eighteen of the wild panthers in Florida have radio collars so that their movement patterns and activity can be tracked on a daily basis. Offspring from some of these animals will be used in the captive-breeding program. It is sad that we have brought the panther to such a point that it requires not only a welfare program but also a life-support system. Once the numbers of panthers in captivity have reached satisfactorily high levels, the USFWS will propose that the Florida panther be reintroduced throughout its historic range in the southeastern United States. But is this realistic? Obviously they cannot live near Disney World or Daytona Beach. Where does suitable habitat still remain in Florida? If the big cats cannot make it in the wild with the habitat now remaining, how are they going to do so in ten years? If land development continues at current rates in Florida and elsewhere in the Southeast, we will have significantly less wild habitat, not more.

Dennis Jordan of the USFWS believes that numerous habitats that could support panther populations still exist throughout the Southeast, including Florida. However, he questions whether people will tolerate them. Unfortunately, his concern is probably a valid one. Two clear options exist. One is to be complacent about the fewer than five dozen panthers that are hanging onto the southern tip of the United States by their last claw. This means we are willing to accept environmental defeat with the panther's ultimate extinction. Another option is to reestablish natural habitat within the panther's former range, especially corridors connecting large, protected areas where the cats formerly lived. We will need a lot of environmental attitude adjustment (or some strong hurricanes) before Floridians give up their housing developments and citrus groves so that panthers can reclaim their home. Reestablishing natural habitat would also require dispelling common fears in southern Florida, or anywhere else where they might be reintroduced, that panthers are dangerous to people or eat too many deer. Neither belief has any factual substance.

The most difficult step would be inculcating acceptance of revised land use regulations. Landowners are quick to rebel against changes in property laws; however, we have reached a point where decisions like this must be faced. And the tough choice is the only option to assure ourselves of a chance to see America's most graceful and majestic cat in the wild.

In contrast to people who persist in believing the mountain lion is still part of the fauna in the Southeast, many people do not realize just how prevalent certain wildlife species once were. With today's wide variety of environmental disruptions, some species have been completely eliminated from one part of their geographic range but not another. Because of such local extinctions, some occurring more than a century ago, present-day residents in a region may not be aware that a species was once present. For example, in the early colonial days, the eastern United States was home for elk. Even buffalo roamed in small herds in the Carolinas, Virginia, and Florida. But the disappearance of these and other species from the East occurred several generations ago, so long ago that many people do not know they were once there, nor yearn for their return.

One species that was once part of our native fauna throughout most of North America is the wolf, known in some parts as the gray wolf or timber wolf. The wolf is clearly a species that has disappeared from many regions because of human attitudes. The following passage by Aldo Leopold describes our typical treatment of any wildlife species we perceive either as a threat or as not contributing directly to our well-being. "In a second we were pumping lead into the pack. When our rifles were empty, the old wolf was down and a pup was dragging a leg into impassable slide rocks. We reached the old wolf in time to watch a fierce fire dying in her eyes."

Our attitude toward the wolf typifies this country's past and, unfortunately, present stance toward some wildlife, particularly species that compete with us in any way. The wolf's shrinking geographic range during the past century is testimony of the effect of this attitude. When the first white colonists arrived, wolves lived throughout North America. By the end of the Civil War, wolves had disappeared from a third of the states east of the Mississippi, primarily heavily settled areas of the Northeast. At the turn of the century, more than eighty thousand wolves were shot, trapped, and poisoned in Montana over a twenty-five-year period. When the troops came home from World War II, wolves were extinct in 80 percent of the eastern states and 33 percent of those in the West. According to Monty Sloan, a wolf behavior specialist with Wolf Park in Battle Ground, Indiana, wild packs are now confirmed in six states, including Alaska with at least five thousand wolves. As many

73

as eighteen hundred live in Minnesota. Fewer than one hundred remain in all of Wisconsin, Michigan, Montana, and Washington.

Why were wolves exterminated from America? Our forefathers saw the wolf as a ruthless killer of livestock, poultry, and even people. Indeed, wolves would have no inhibitions about attacking and eating calves or chickens, but try finding a noncontroversial record of a wolf killing a human in North America. It won't be easy. Wolves are normally very shy toward humans. One wolf authority who actually had two with him as pets when we talked, claims there is no record of a full-blooded North American wolf even biting a human. Other wolf experts disagree, but when cases are this difficult to document, such events are certainly rare under natural conditions. Nonetheless, tales of wolf violence are told, retold, and believed. Stories such as Little Red Riding Hood and the Three Little Pigs probably have not helped the wolf's reputation with the younger set.

Wolves were seen as powerful competitors in a developing land where survival odds were already steep. To the wolf's misfortune, its family style contributed to its downfall. Wolf pairs sometimes mate for life, maintaining a close-knit, loyal family group. This enabled bounty hunters to kill one or both parents and all of the young at one time. The last bounty for wolves (thirty-five dollars per wolf in Minnesota) was not removed until the mid-1960s.

A large male wolf normally weighs from 70 to 100 pounds and can weigh more than 150 pounds. Wolves sometimes travel more than a dozen miles daily in search of food, and a few individuals have been documented to have home ranges of almost two thousand square miles. The wolves' major large preys include sick or wounded deer and, in the far North, caribou. Populations of smaller preys such as rabbits and rats are kept in check when wolves are around. When wolves kill domestic animals, it is often because humans have created environments where their normal preys are scarce.

But we pay a price each time we find ourselves unable to coexist with another species. Part of that price is the cultivation of an environmental attitude that rationalizes our elimination of any plant or animal competitor that gets in our way. Such environmental intolerance is no longer acceptable. Do we really want to eliminate large, exciting animals like wolves simply because they may cause a minor problem once in a while? Particularly since they normally pose no direct major threat to humans. Wherever they live today, we should remember that they were there before us.

Wolves represent a segment of our wildlife, the predators, that is never as abundant as typical game species such as deer, rabbits, and quail. Yet they have a distinct and important role in shaping the wildlife community. And they represent an untamed wildness that lends strength and vigor to our natural heritage. Destroying such animals as wolves, panthers, and grizzly bears is like removing the roller coasters from amusement parks. Some of the thrill disappears. Leopold, a famous conservationist of the early 1900s, expressed the proper sentiment at the end of his description of killing the mother wolf and her pups: "I realized then . . . that there was something new to me in those eyes—something known only to her and to the mountain. . . . I thought that because fewer wolves meant more deer that no wolves would mean hunters' paradise. But after seeing that green fire die, I sensed that neither the wolf nor the mountain would agree."

Some farmers and others who spend their time outdoors accept that humans can adjust their lives to coexist with other animals. However, public attitudes about wildlife issues are often emotional responses provoked by specific incidents or by misinformation. Often, all the biological facts may not be considered.

Another animal that has been greatly affected by human attitudes is North America's biggest reptile, the American alligator. Everyone knows that enormous reptiles called alligators live in southern lakes and swamps, and have big teeth and powerful tails. Alligators get people's attention—through fear or fascination. (I often wonder what's more interesting, the alligators themselves or people's attitudes toward them.) Numerous legends and tall tales can be expected to develop about a reptile that reaches a length of more than thirteen feet and can kill and eat a full-grown deer. Some kernel of truth may even be found in such stories.

The American alligator occurs throughout much of the coastal southeastern United States to a hundred miles, or even farther, inland. Of the fewer than two dozen species of crocodilians remaining in natural habitats in the world today, alligators occur farther north than any other.

Will alligators attack humans? Under normal conditions, the American alligator will not. Rare but spectacular reports of seemingly unprovoked attacks by alligators on humans keep alive an awareness that alligators can be highly dangerous. This blemished record for good behavior has been further tarnished by another trait, a strong maternal instinct to protect the young. Mother alligators should receive a lot of credit for proper parenting. They do a better job than some people.

All crocodilians lay eggs. The female alligator builds a large nest of mud and

75

vegetation along the shore and deposits approximately thirty to sixty white eggs in early summer. Young hatch in the late summer or fall and enter the water at that time. When the young hatch, the female may crawl up onto land, open the nest, and gently carry the babies to the water in her mouth. If necessary, she will crack the hard-shelled eggs in her mouth so the babies can escape unharmed. Because of this powerful protective instinct, a mother alligator often remains in the vicinity of the nest. Thus a person approaching the nest area may suddenly find an enormous, hissing reptile charging overland. Unless you molest the nest or pick up a baby, she may retreat if you stand your ground.

Alligators have a problem faced by a lot of us—you can pick your friends but not your relatives. Not that alligators have many friends, but they do have some notorious kin. Included among the crocodilians are the few species of animals left on earth that will attack a full-grown human for food. Based on the evidence, American alligators do not consider humans a menu item. But as close relatives of the Old World crocodiles, some of which are uncontested man-eaters with awesome reputations, they will never be completely above suspicion. An unmolested alligator is shy and peaceful, but true tales of crocodiles eating humans in Africa and Indonesia do not help public opinion about their New World counterparts. The American alligator exemplifies how a species can have its natural rights violated because of public misunderstanding.

Should we be sympathetic toward someone who risks finding out what a watchful mother alligator will do to defend her young? Is it fair to condemn an alligator that has lived for thirty years in a coastal swamp, has watched its habitat be converted into a resort, and then proceeds to eat the first poodle that walks down to the edge of the new golf course? Some people, especially poodle owners, might say the alligator has no right to do these things. Others would disagree. The alligator is simply defending its young or eating a small prey animal as it and its ancestors have always done. The dog owner should be more educated and more responsible, as should the baby alligator molester. We should be more accepting of the alligator's place in the natural world, remembering that the alligator was here before us by about two hundred million years.

Alligators have their place in the world of nature and we have ours. But sometimes our paths cross. Should we become indignant that an alligator living in an aquatic area where people are advised not to swim bites a swimmer? Why should anyone, other than the victim, take responsibility for an alligator attack in such an area? An attack on a child is a different matter, but even then the alligator should not automatically be indicted as the bad guy. After all, who let the child go swimming or wading with alligators? Who provided the marshmallows to feed the alligator down by the lake with the sign that warned "Do Not Feed the Alligators"?

A few simple rules for people to observe about alligators are based on common sense and my limited background in alligator ecology. First, any big animal with teeth can hurt you. Therefore, do not feed, tease, or even approach alligators in the wild. Second, alligators are extremely protective of their eggs and babies. Do not go near areas where they may have nests; and never pick up a baby alligator, no matter how cute it is. If you live in or visit a place with alligators, make sure your children know and follow these rules just as they follow the rule of looking both ways before crossing the street. Living with alligators is simple if we don't try to become a part of their world or encourage them to become part of ours.

Some people in southern states, particularly Louisiana and Florida, express wonder at the need to continue protecting American alligators, which appear to be in overwhelming profusion in many areas nowadays. But how abundant were they only three decades ago when alligator poachers were making their claims? Not very. Southern history clearly shows that alligators practically disappeared from the region after a few years of assault by hide-seeking poachers. If you are glad that we have, so far, avoided driving this species to extinction, if you appreciate having these creatures back on the environmental scene in the South, do not be fooled by the apparent abundance of alligators.

Federal and state laws are already beginning to relax. Some states have opened hunting seasons, promoting a perception that the species does not need much protection. Alligators are very easy to kill, and if they again become marketable on a broad scale, for their hides and perhaps their meat, they will have little chance of survival. Once their slaughter is permitted on a widespread basis, illegal poaching could occur in unlicensed areas, and untold numbers might once again be killed. If commercialization can be prevented so that wild populations are protected, the American alligator will probably remain as a most unusual part of our natural heritage.

Alligators are an environmental legacy we do not want to lose. They never cause property damage and seldom threaten people. They are also a true keystone species.

Alligators hold the franchise on many wetlands within their geographic range, and many species of animals depend on a certain aspect of their behavior. Alligators dig large depressions, known as "gator holes," in the bottom of lakes. James Kushlan of the University of Mississippi conducted a study in the Everglades during a drought, when lakes and ponds dried up. He showed that the only open surface water remaining in some areas was that in the gator holes. These aquatic sanctuaries were critical to the survival of numerous fishes, reptiles, and amphibians. Of course, the alligators picked up a few easy meals, but their role in creating gator holes was essential to the survival of individuals of some species.

Alligators are also a benefit to bird life in some southern ecosystems. When heron and egret rookeries are on islands whose surrounding waters are patrolled by alligators, they are less likely to be raided by certain predators. Raccoons and opossums cannot safely enter waters where big alligators live and are therefore discouraged from swimming out to the nests to eat eggs or young. Ultimately, the reptiles profit from their role as guardian, and it is adult wading birds that pay the price as one of the preferred preys of alligators.

Although the eleventh hour has come for panthers in the East, and midnight has tolled for wolves, time has not yet run out for alligators. They will continue to persist if we will but let them. It's all in our attitude, including an acceptance that they are a part of the natural habitats of the southeastern region. We must also learn to accept the natural behavior of all animals. Alligators will eat dogs. They are potentially dangerous to humans. Our acceptance of the way they are, and a modification in our own behavior in return, is all that is required. The protection of the American alligator, or of any species that is a potential threat to us, lends credence to the assertion that we are truly dedicated to preserving our natural heritage.

Don't Tread on Me

Nearly everyone has an opinion about snakes, and the usual result of those opinions is unfortunate for the crawling reptiles. Thus snakes are a sensitive barometer of environmental attitudes. Snakes are among my favorite animals. I continue to wish more people would appreciate them. Not as pets or members of the household but as important components of our natural environments. People are gaining a greater awareness of and concern about the welfare of natural environments and wildlife. This emerging philosophy should include our snakes.

Most people vividly remember their encounters with snakes. As a professional herpetologist (one who studies snakes, other reptiles, and amphibians), I have had more opportunities than most people to observe snakes. One memorable occasion was the week my son, Michael, and I encountered thirty-eight rattlesnakes of nine different species. The place was Arizona. Because Phoenix is fast losing its natural desert charm and was hot, very hot, we headed southeast toward the cool Chiricahua Mountains. Our guide for the week was Randy Babb, a friend who lives in Phoenix. Randy is a rugged, likable, nature-knowledgeable individual who epitomizes an environmental outlook that we need more of in this country—an enthusiasm to see, touch, hear, smell, and taste nature, with the idea that it should all be there the next day for someone else. When Randy turns over a rock to see if a snake is under it, he always puts the rock back in the exact same spot.

Our general goal was to experience the habitats, animals, and plants of the area.

I wanted to learn more about the herpetofauna of the Southwest, Mike wanted to see a lot of rattlesnakes, and Randy just wanted us to have a good time. Everyone's wishes were granted. The American Southwest still harbors some exciting country, in addition to its rattlesnakes. We saw giant saguaro cacti, jackrabbits, and flash floods. At night we heard coyotes howling in the darkness around our camp. If you like wildlife and natural phenomena, you would have enjoyed the trip.

80   Arizona now has strict laws against collecting animals. Permits are required, even for catching snakes, and the sale of native animals is forbidden. Although some courts are still perceived as being lenient toward the illegal pet trade, Arizona has made a strong environmental statement about the preservation of natural environments and the plants and animals that inhabit them.

One day we drove to a mountain site where Randy said we might find twin-spotted rattlesnakes, a rare, state-protected species that cannot legally be harmed in any way. None of us had ever seen one in the wild. The four-wheel-drive jeep ride took an hour on a road that looked (and felt) like a dried-up, rocky stream bed. At eighty-two hundred feet we left the jeep and climbed another two hundred feet up along rock ledges. Bears and mountain lions, which use such ledges, had been sighted in the area recently. Not surprisingly, we were the only people around. Twin-spotted rattlesnakes are small, and this one was coiled on a ledge that I had managed to pull myself up on. I was about two feet from it when I heard the high-pitched whir of its warning rattle. Fortunately I did not follow my first impulse: to jump from the ledge. We were probably the only people anywhere that day to see a twin-spotted rattlesnake in the wild.

Some may wonder why we pursue such activities so enthusiastically. Gary Meffe, from the Savannah River Ecology Laboratory, who joined us on this trip, put it this way: "Is it any surprise that people sometimes wonder about ecologists, particularly herpetologists? Imagine my telling my neighbors that I flew two thousand miles to Phoenix, drove another two hundred miles to some mountains along the Mexican border, bounced and bumped ten miles up a road that would destroy a normal car, then climbed a rock cliff from which you could fall to your death, in search of a rattlesnake that could kill you if it bit you." This is probably not everyone's idea of an ideal vacation or a properly planned field trip. Rattlesnakes are not most folks' chosen brand of wildlife experience. But for those who wish to get out and meet Nature on her own terms, snakes can play an integral part in the experience.

A herpetologist on a quest for a particular snake species is akin to the fisherman after rainbow trout, the hunter after elk, or the bird-watcher looking for painted buntings.

No matter where you live in North America, spring is the time of year you will most likely have a close encounter with a snake. You may not actually see one, but you can be almost certain that some will be aware of your presence. Usually you have no cause for alarm. The reason the snake did not inform you of its presence is the same reason other animals keep quiet about their existence. They don't want to hurt or bother you—they just want to be left alone.

Certainly snakebite is a possible hazard of being outside, as are cuts from broken bottles and puncture wounds from rusty nails. But don't let fear and ignorance keep you from enjoying the outdoors. You are safer in the woods taking your chances with the snakes than you are on the highway you travel to get there. The last thing a snake, even a rattlesnake or cottonmouth, wants to do is bite an animal it can't eat, especially one as big as a human. Biting is normally the last resort, when escape seems impossible or the threat seems extreme, for it can result in broken fangs and lost venom needed to capture prey. If a rattlesnake can keep you away by vibrating its tail, it is better off than if it has to bite you.

Because of peoples' fascination with and fear of snakes, they serve as a barometer of the public's sensitivity toward wildlife and natural habitats. Attitudes about snakes are a measure of the extent of environmental education in a region. One way to gauge people's knowledge about snakes is to find out what they think is true or false. Remember the coachwhip snake that was capable of outrunning a man, usually a sinner, and then whipping him to death? The snake would occasionally stop the whipping and hold the end of its tail below the victim's nose to see if he was still breathing. Unfortunately, some people still think this is true. A coachwhip does indeed look like a braided whip, can reach a length of seven feet, and appears to move fast. But one will not use its tail to whip you or determine if you are still alive. The nonvenomous coachwhip can only give you a harmless bite.

One modern untruth about snakes is that you can keep them out of your yard by sprinkling sulfur around it. I do not know what sulfur might do for your plants, but I know of no scientific evidence that a ring of sulfur or any other store-bought chemical will make the slightest difference to a snake. Of course, a snake will probably not want to coil up on a bed of yellow powder if given a choice. But by the time you put out enough sulfur to discourage the snakes, you will have probably discouraged any houseguests and other visitors, including the letter carrier. Nonetheless, people in some areas still buy sulfur to ward off snakes.

Another yarn I hear from time to time is that snakes chase people. No venomous snake in this country will intentionally pursue a person. True, some will stand their ground and even strike out at you if you get too close, but they don't ever come looking for you. If someone in your region thinks he or she has been chased by a snake, it's almost certainly a misperception. Perhaps the snake crawled in the person's direction because that's where it was headed, unaware of the human presence.

82

Being professional herpetologists, my associates and I have encountered more than twenty thousand snakes, from California to Canada to North Carolina, and none of us has ever seen a venomous American snake chase someone. I qualify this with "American," because some of the African mambas and Asian cobras are said to take offense at a person's presence and actively attack. Maybe they will, but for the species in this country, not so.

One myth that seems to have been perpetuated about the time waterskiing became popular is the one about the skier who fell and landed in a bed of water moccasins. I have been told this story dozens of times. When I ask where it happened, the answer is that it was reported in the newspaper. When I ask what newspaper, the answer is that someone else told them about it being in the newspaper. I have yet to find that newspaper article, but then again, I don't buy the ones in the grocery store checkout line.

A while back I did see a newspaper column in which someone was defending a policeman released from duty for shooting a snake. As I remember the story, the snake was harmless (a king snake, I think), and the columnist was supporting the attitude that snakes should be killed on sight—an attitude based on some kind of nonlogic like "Shoot first, and then find out what kind of snake it is." Such an article may have been appropriate in colonial times when explorers and settlers were ignorant of what perils awaited them in the wilderness. But the attitude is senseless in the last decade of the twentieth century, when wildlife is disappearing at an ever-accelerating rate. Our natural environments, which include snakes, are priceless. The simplest rule for those who don't like snakes is to leave them alone. We should be teaching wildlife facts, not promoting the continuation of ignorance and irrational fears. I don't tell people they should never kill a snake. Clearly, in some situations it might be advisable. But the capricious destruction of any native wildlife is irresponsible. With the increase in our level of education, today's public should know (though not enough do) that the majority of snakes in the United States are not dangerous. Only seven of more than fifty snake species in eastern North America are venomous.

Snakes still have more myths surrounding them than does any other group of animals, but the days of bizarre superstitions are over for most people. The public has clearly become more accepting of snakes than it once was. Even the act of intentionally running over snakes on the road is less common, although occasional skid marks that swerve toward a dead snake on a highway indicate that we still have a few insensitive drivers left.

83

Although injury or even death from snakebite is a possibility, the number of lethal accidents from household cleaning fluids, electrical wiring, or lightning far outnumber those from venomous snakes in North America. We have little to fear from most snakes in this country, but I do know of a few instances of a snake striking back at someone who would have been better off had he left the snake alone. Steve Bennett of the South Carolina Wildlife and Marine Resources Department collects snake tales and related to me three stories that give evidence that attempted destruction of snakes can have dire consequences—and at least three people should have curbed their distaste for these reptiles.

One story was of a man who was armed with a rifle when he encountered a snake. Apparently not wanting to waste ammunition on such a lowly creature, but nonetheless wanting to kill it, the man began crushing the snake's head with the butt of his gun. Anyone who has seen a dying snake knows it begins to thrash and curl around. As the snake squirmed, its tail reached the trigger and squeezed. As far as we can determine, this is the first instance of a snake shooting a man!

In another incident, a man with a double-barreled shotgun saw a snake crawling around one of his outbuildings. As the snake slid alongside a box in the doorway, the man pulled the trigger of the shotgun. He never pulled the trigger a second time, but not simply because the snake was dead. The man was dead, too, and the building was gone. The first shot had detonated the case of dynamite the snake was crawling beside.

Some readers may feel the final story has a happier ending because no people were killed. Believing that snakes infested her mobile home, a woman decided to try sulfur as a snake repellent. But this woman decided to go beyond just sprinkling sulfur around the yard. Instead, she filled pots with sulfur, placed them under the trailer, and then set them afire. Now powdered sulfur doesn't smell particularly good, but burning sulfur smells terrible, like rotten eggs. I doubt if snakes care much one way or another about the smell of rotten eggs. And if the woman had not accidentally burned the trailer to the ground, the snakes probably would have continued their residency (though the human residents would certainly have moved out because of the odor). But at least she got rid of the snakes.

Pondering the stories recounted above may lead some people to believe we have pushed our luck with snakes about as far as we should. Once we conquer our adverse attitudes toward snakes, which are as much a part of our natural heritage as bluebirds, bobcats, and yellow jessamine, we will have reached a pinnacle of environmental acceptance.

We are not there yet. I recently ran across another example of a fearful, unenlightened attitude toward snakes. The person in charge of a major safety program for an enormous company decided to ban the use of live snakes in presentations to its employees, each of whom are required to attend a monthly safety meeting. The reason given was that the presence of snakes in a room with people was in itself a safety hazard, even though many people in the South and West could benefit from a talk on snakebite safety. What better way to allay ignorance, which is more than half the problem with snakebite, than to use live snakes? Would you attempt to teach chain saw safety but prohibit a presentation with a chain saw?

Obviously, there are still some major improvements to be made in the general public's attitude toward snakes. Further evidence of this is seen in the form of a most unusual questionnaire I received from a college student, Len West, in Wichita Falls, Texas. She was doing a class project that involved getting public opinion on the effects of rattlesnake roundups. A well-prepared questionnaire, as this one was, is instructive to those filling it out by making them ponder an issue. A majority of people who enjoy being outdoors presumably appreciate wildlife and have a respect for nature, but occasionally nature gets a slap in the face. The rattlesnake roundup is an environmental insult that might be considered a punch in the jaw.

Here is a quick review of what can happen at an unregulated rattlesnake roundup. The participants gather in a region with an abundance of rattlesnakes and a community environmental IQ well below normal. As in a fishing rodeo, individuals compete for recognition by capturing the largest or the most. Financial profits can be made based on the number of pounds of rattlesnakes sold to buyers who attend the festival. But in contrast to fishing rodeos, some rattlesnake roundups still have no guidelines and no controls. No one needs a special permit, although a hunting license may be required in some states, and no one has a creel limit. The people of a region simply have part of their natural wildlife destroyed or removed. The participant pays no fee. And rattlesnakes are not the only wildlife affected.

Many roundups are held when snakes are hibernating underground in burrows or rock crevices. Among the capture techniques is "gassing," which is now illegal in some states. Environmentally responsible rattlesnake collectors do not use the technique, and some buyers refuse to purchase snakes collected in this manner. Gasoline fumes are forced into a burrow or crevice through a plastic hose. Out come the inhabitants. Or at least most of them. Some die while still in the burrow and those that crawl out often die later. Not only rattlesnakes are affected. Depending on the location of the gassing, a rattlesnake roundupper might drive out or kill indigo snakes (a federally protected species), gopher tortoises (for which several southern states are trying to enact stronger protection laws), and a host of other animals that have sought sanctuary in the hole. The snakes captured at a rattlesnake roundup may be sold or bartered as pets, or for skins, meat, or curios. Or they may be released in the vicinity or somewhere else, thus affecting another natural habitat.

One justification originally given for rattlesnake roundups might have been acceptable in biblical times. Snakes, thought to be evil, were meant to be destroyed. Such ancient ideas are certainly still around. Another justification is that venomous snakes should be removed because of the danger of snakebite to humans and livestock. This might be acceptable in parts of Asia where deaths from snakebite are in the thousands per year. But in the United States almost as many people are killed each year by turning soft-drink machines over onto themselves as are killed by rattlesnakes. I doubt if rattlesnakes kill many quarter-ton cows either, although they almost certainly cause more deaths to cattle than does the deadly soft-drink machine.

Although the snakes and other wildlife suffer during rattlesnake roundups, some rural communities may gain economic benefits. Tourists are attracted, charities are supported, and local businesses do well for a few days. Whether such events can be sustained with local populations of rattlesnakes is unlikely. The argument that snakes are just part of the environment, which belongs to the people, and are thus fair game for people to treat as they please is weak ground to crawl on. Unconstrained

85

exploitation of the environment and native wildlife is neither feasible nor acceptable. Rattlesnakes and other wildlife species belong to all of us. And not all of us approve of the uncontrolled, unregulated destruction of them.

But rattlesnake rounduppers are not doing anything illegal in states that have no regulations protecting rattlesnakes. A free license to exploit native wildlife reflects poorly on the people who live and vote in states with unregulated rattlesnake roundups. In an environmentally enlightened region, the voters would not allow such an activity to go uncontrolled. Not only would regulation and control of rattlesnake roundups be ecologically sound, it would be good economics. Any short-term economic benefits enjoyed by a community that permits unrestricted rattlesnake roundups are far outweighed by the long-term disadvantages. Why should the citizens of a state allow exploitation of a wildlife species without enforcing some rules and charging a fee? Why should we allow environmental destruction of any sort without having done the research to determine the overall impact on other animals as well as the target species?

As with other environmental abuses of today, the real culprit turns out to be those of us who have not put the laws in place that provide the necessary environmental protection. Obviously, we still have some environmental educating to do, particularly in the snake arena. Every plant and every animal has a right to exist in its natural habitat. We are fast destroying the natural habitats of the world. And we cannot, and should not want to, exist without them. Rattlesnakes and rattlesnake habitats were here several million years before us.

Snakes fascinate almost everyone, either through nescience and fear or through simple curiosity about an animal with an unusual life-style. To understand the natural behavior and ecology of snakes, you have to spend a lot of time studying them and observing them in the wild. Even the most knowledgeable snake enthusiasts realize they still have a lot to learn. But snakes, along with the many other reptiles inhabiting North America, are faring poorly in the face of urbanization, agricultural expansion, and some forestry practices. The draining of wetlands and other forms of habitat destruction obliterate suitable living places for many of them. Some, such as garter snakes, which can survive even in cities, still reside in wooded lots and backyards across the country. But other species are in big trouble. One in particular deserves special attention.

Eastern diamondback rattlesnakes are the largest species of venomous snake in the United States. They normally have no problem with food supply. A rat that runs in front of a hungry rattler becomes a fast-food meal. And from natural predators adult eastern diamondbacks have little to fear. They have a formidable defensive attitude when confronted and are capable of delivering a deadly dose of venom to a full-grown, healthy adult. Yet diamondbacks stand no chance against the attitudes and habits of humans. The diamondback's ability to defend itself against any intruder has rung the death knell for this species. Few people, especially legislators, are willing to speak up in favor of an animal that can kill you.

Diamondback rattlesnakes are popular with a few people, mainly snake collectors who enjoy the ultimate thrill of confronting a truly dangerous animal on its own turf. Such people are adventurers, willing to accept the consequences of entering a diamondback's habitat. It's the snake's home, and it may not like your being there.

People who have had pets, livestock, or relatives bitten or killed by a diamondback or other venomous snake may be unsympathetic to recommendations that they be protected. But whereas more people have been killed by dogs than by eastern diamondbacks, they usually condemn the particular dog, not the whole species.

I recently read an article written by a first-rate field naturalist, Heyward F. Clamp, Jr. He has probably caught more eastern diamondback rattlesnakes than I've ever seen. A native of South Carolina, he once caught rattlesnakes for a living. Now he fears an end is in sight for the species. He still searches for them as a pastime, the same as a bird enthusiast looks for birds in a forest. But no longer does he remove large numbers from where they belong. To his dismay (and mine), the large numbers can no longer be found in most places. Too much environmental destruction has occurred through uncontrolled development of the diamondback's habitat.

Heyward favors protection for this most impressive reptile. Ironically, a species that can be lethal when it protects itself from natural enemies must be singled out for special protection from humans. This is a difficult concept for some to accept, but the time has come to give the idea of protecting dangerous animals thoughtful consideration. We stand to lose this majestic species, and many others that are in danger, simply because they dare defend themselves. As Heyward says, "Like the early Americans who could see no end to the buffalo on the plains, like the plume hunters at the turn of the century who could see no end to the tens upon tens of thousands of egrets in the Florida Everglades, we thought the diamondbacks would be there forever. And in our innocence we took hundreds of these magnificent animals from the South Carolina Coastal Plain."

Snakes, along with all other animals and plants, are part of our natural heritage. The rattlesnake emblazoned on Revolutionary War flags represented vigilance. The enemy was the British; the country was at stake. Today the rattlesnake is a symbol of environmental vigilance. The country is still at stake.

Are we reaching a more progressive stage of environmental awareness? The collecting and killing festivals known as rattlesnake roundups are being challenged. Snakes are being placed on the federal endangered species list. And most snakes I am asked to identify these days are alive instead of in several pieces. I like to think we are learning to accept all of the native wildlife inhabitants of the world.

And the Good News Is
Franklinia

The elimination of local populations of plants and animals, often a precursor to extinction, usually results from human intervention. Ironically, our treatment of certain species and habitats has actually resulted in their unintentional preservation. One example is the tree *Franklinia*, one of the most mysterious plant species ever found in North America.

Proper management of our environment is a controversial issue. Some maintain that environmental management is an important exercise in preserving natural environments. Others argue strongly against any human involvement. Amid today's turmoil and controversy about how to manage our environment properly, the fate of *Franklinia* strikes a positive note. The story of this plant serves as a reminder that human intervention into nature's ways can sometimes have a happy ending.

Franklinia, like its relatives sasanqua and the camellias, is in the tea family and can be purchased from horticultural companies. But if you discover it growing in the wild, you will make botanical history. For *Franklinia* is not simply a variety of the original plant that is now produced in horticultural greenhouses; it is the last representative of a group of plants, an entire genus, that has disappeared from the wild within our country's lifetime.

In the late 1700s, William Bartram traveled widely throughout the Carolinas, Georgia, and Florida. He kept careful records of his natural history findings. In the autumn of 1765, he discovered a new plant that blossomed in fall and winter instead of in spring. He reported a small grove of trees with white flowers growing along the southern reaches of the Altamaha River in Georgia (hence, the plant's scientific name, *Franklinia alatamaha,* after Benjamin Franklin and an earlier spelling of the river where the plants were found). Bartram did not encounter the plant anywhere else. By the time he returned in 1773, the species had become recognized as rare and unusual by botanists, so Bartram collected seeds and sent them to England for identification.

90

Then came the mystery. Since Bartram's visit and those of a few other explorers who also found the grove in the late 1700s, no one has been able to find the site of the *Franklinia* trees. The plant apparently became extinct between the time of Bartram's travels and 1800. Another species gone forever. But wait. What about those seeds that Bartram collected? As it turned out, they ended up in Philadelphia instead of England. But the seeds were viable! *Franklinia* seedlings soon began to sprout. Although no longer found in the wild, *Franklinia* can still be found through-out the country, surviving only as a cultivated plant.

We can only speculate on the cause of disappearance of the species from southern Georgia. One explanation given is that *Franklinia* was not actually native to North America but was brought to the New World by English tea growers trying to start a new crop. Bartram may have stumbled onto the last remnants of an unsuccessful crop venture abandoned by early settlers. Another suggestion is that the plant, from tropical American rain forests of high elevation, was brought to the Southeast by Indians or European explorers. Indeed, cultivated *Franklinia* do seem to thrive in the cooler mid-latitudes of North America. However, neither explanation is totally satisfactory. The question of where the plants are in the wild is still unanswered. Botanists cannot find them anywhere, except for those raised by horti-culturists.

One possible solution to the *Franklinia* puzzle may be that Bartram discovered the last grove of a nearly extinct species of native plant. Perhaps it was an undescribed species, destined for extinction in obscurity. Or the trees in southern Georgia may have been a relict population—the last survivors of a once widespread plant species. In any case, as a consequence of natural environmental events such as changing climatic conditions or the plant's susceptibility to an insect or blight, *Franklinia* was almost extinct by the eighteenth century. Presumably *Franklinia* is gone from the wild, but who knows? It is an exciting possibility to consider that, within 150 miles of Atlanta, there still exist swamps and river bottoms that are rugged and wild enough to hide the last remnants of a rare species that almost went extinct.

Franklinia is not the only plant presumed extinct in its natural habitat in the wild but that nevertheless manages to persist. Two other prominent examples are a tree that smells awful and a shrub so well known that a song has been written about it. The tree is the ginkgo, with a historical botanical record that traces it back to temple gardens of China. Yet, ginkgoes are apparently extinct in the wild, and no one knows for sure where they came from.

Only one species of ginkgo tree exists, and it is the only surviving member in the family Ginkgoaceae of the order Ginkgoales. To put this in perspective, all oak, chestnut, and beech trees belong to a single family, the Fagaceae. The family Fagaceae, along with the trees in the birch family, constitute the order Fagales. The ginkgo is the sole survivor of a major family in a major order of trees whose fossil record extends back to the Paleozoic era, more than 230 million years ago. The ginkgo is sometimes referred to as a "living fossil" because in natural environments the only record of its past is in the form of fossils. Indeed, it has one of the longest fossil records among the higher plants. Although it has no confirmed wild populations, it prospers today in the Orient and North America, almost always within planting distance of a house or building.

Although the fan-shaped leaves are broad and flat, like hardwoods, ginkgoes and their ancestors are more closely related to conifers. The two sexes are distinct, with a particular tree being either male or female. The trees can reach a height of more than eighty feet, and the leaves turn bright yellow in the fall. Ginkgoes are nice trees to have around, if they are males.

Female trees are attractive to look at in the fall, but their visual beauty wanes with the impact the fruit has on another sense—smell. One who is unfamiliar with ginkgo trees and passes near a fruiting female instinctively lifts a foot, casting a suspicious eye for unleashed dogs. The smell of the ginkgo fruit has been described in more delicate ways. Authorities have referred to the odor as one of "rancid butter" or just an "unpleasant smell." The smelly, yellowish fruit looks like a plum or apricot, and some adventurous botanists claim that the enclosed nut tastes good if roasted.

So far as I can discern, no one knows why ginkgo trees disappeared from the wild in the Orient. They are pollinated by the wind, so we cannot assume that their demise was because some rare pollinator or seed disperser became extinct. Was this a type of tree that was overused as a major wood product by some ancient Oriental culture? Most likely, the ginkgo tree was like *Franklinia*. Only a few surviving members of the species existed as a result of competition with other species, a pervasive blight of some sort, or a specialized predator. Interestingly, however, ginkgoes are noted for their resistance to fungus, fire, herbaceous insects, and industrial pollution.

No one knows why ginkgoes survived in those Chinese temple grounds, although we can surmise that some religious order chose to save seeds and replant them. Whatever the case, ginkgoes are here with us today, despite their official ecological disappearance centuries ago.

To Americans, perhaps the best-known plant without a home is the star of the song "Tumbling Tumbleweed." Tumbleweed is as much a part of the legends and movies of the Old West as stagecoaches, cowboys, and cattle drives. But in one sense, tumbleweed did not really belong there. In the early 1870s, the ancestors of today's plants were probably still across the Atlantic. The plant, known by botanists as Russian thistle, was apparently introduced by accident from Europe in a shipment of flax seeds, though the species is not native to Europe either. Tumbleweed received its first attention in the New World from farmers in the Dakotas, where it was viewed as a pest.

Tumbleweed looks good in a Hollywood western, rolling across the path of a posse on a dark and windy day. But the leaves are sharp, and people and horses alike can suffer painful cuts from the spines. In the modern-day West, a collision with tumbleweed can also result in the need for a new paint job on a pickup truck. In earlier times, the potential devastation from prairie fires was increased dramatically by the prospect of a tumbling fireball, which might eventually come to rest against a building. The huge, dried balls of plants added to their nuisance value as a fire hazard by piling up against houses, barns, and fences. The agricultural community of the Great Plains felt the brunt, as the rolling plants spread their seeds across the land, invading croplands, towns, and open prairie.

Tumbleweed was so destructive on the American frontier that some farmers even hinted at conspiracy. Russian Mennonites, an identifiable social and religious group with customs different from local ones, were already a target for frontier discrimination. As tumbleweed problems became more intolerable, some farmers tried to place the blame on the Mennonites. They claimed that the religious group had intentionally introduced the noxious weed to the Dakotas in a vindictive plot to get even for the prejudices against them. The U.S. secretary of agriculture sent a botanist to the region to study the tumbleweed problem biologically and to seek control measures. He eventually demonstrated that the introduction had been accidental and dismissed the claims against the Mennonites.

Ironically, tumbleweeds owe their success to agricultural practices. A tumbleweed plant disperses its seeds, up to a quarter of a million per plant, by rolling across the landscape. The clearing of forests that created open habitat and eliminated tall prairie grasses aided such a way of life. Tumbleweeds soon disappeared in dense grassland habitats that did not permit suitable dispersal of seeds or seedling establishment.

Tumbleweed is a botanical success story in barren habitats throughout much of the world, including Eurasia, Australia, and North America. However, according to James A. Young of the U.S. Department of Agriculture, "The native population of plants from which the tumbleweed arose no longer exists in Eurasia," despite its abundance in some other areas. The worldwide prosperity of Russian thistle can be credited to an agricultural society that set up ideal conditions for the plant's survival. Perhaps, like *Franklinia* and ginkgo trees, were it not for humans, tumbleweed might be one more extinct species.

We have come down pretty hard on ourselves in the last few years for abusing our environment. And well we should. But we should remember that extinction is ordinarily a natural process. As we know from dinosaur fossils, countless species became extinct long before we arrived. Presumably, each was faced with environmental and population problems that could not be overcome.

Despite the twists with *Franklinia*, ginkgo trees, or the fiction-that-may-come-true in Michael Crichton's novel *Jurassic Park*, in which dinosaurs are resurrected from genetic material, we must accept reality. Under ordinary circumstances we will never be contemporaneous with species brought to extinction in the wild. We may feel a certain sadness that a particular species will never be seen again, but we need not mourn the loss of those that came before us. Instead, we must develop a vigorous appreciation for those species that happen to be here at the same time as the human species and that share the earth with us.

Although tumbleweed might survive without us, ginkgo trees and *Franklinia* are living proof that some species cannot. In many ways this is true for most species on earth today. They need our help if they are going to survive. Ecologists reveal fascinating facts about the lives of organisms on earth today, often with painstaking studies. Sometimes the intricate details of a behavior pattern or an environmental dependency of a species can provide insight into how our manipulation of a natural habitat could cause serious disruptions that would lead to species extirpation—and how a different manipulation would not.

PART THREE

The Search for Environmental Culprits

Who are today's environmental miscreants, the culprits of environmental abuse? Some are obvious: those who exploit natural resources without regard for future (or present) consequences. But not all environmental offenders are so clearly identified. The spectrum of abuse ranges from oil spills to highway litter, from tropical rain forest destruction to lawn and garden pesticides. It's not always easy to say who is to blame. To some degree we all share the blame for the disappearing of the natural world.

Nor is it always simple to determine the most appropriate solution to environmental problems. The conflict between exploiters and protectors of the environment often concerns social and economic impacts that cannot (and should not) be casually dismissed. The perplexity of the problem is immense, and the steady increase in the world's population compounds the problem daily.

The first step in finding appropriate solutions is to acknowledge the problem. The next step is to identify the attitudes and actions that precipitate environmental conflicts. The evolutionary and cultural history of humans has not instilled in us an awareness of the fragility of our natural environment, or sensitivity to the rights of other species, or an understanding of just how limited our natural resources are. If we are to assess responsibility for today's environmental crisis—and, more important, rectify the situation—we must develop that awareness, sensitivity, and understanding.

Not many years ago, attitudes about environmental protection received less than enthusiastic support from most political quarters. The proponents of environmental protection were in the minority, but they fought uphill battles to win concessions that our descendants will one day appreciate.

A lack of political emphasis on environmental issues reflects an attitude that habitat or wildlife protection should not stand in the way of commercial development and progress. When the profit makers hold the political power, "progress" translates to financial profit for a few. But think about it. Is the destruction of what is left of our forests, wetlands, and wildlife really progress? Progress is defined as "steady improvement." Will the next generation of Americans consider endless asphalt roads, acres of shopping malls, and polluted waters an improvement over natural environments? They are far more likely to appreciate that a possibility still exists for them to see a bobcat, hear a screech owl, or smell a yellow trillium in the wild.

We Probably Killed the Last Mammoth, and the Tigers Better Watch Out

Mankind probably began to make significant environmental impacts at least seventy thousand years ago. The control of fire and the development of tools and weapons meant our ancestors could meet the basic needs of obtaining food and defending against predators. The building of boats meant access to habitats in unexplored regions across open water. As humans tamed and explored their environment, the world and its resources were no longer limitless. A new and crafty consumer was emerging, one from whom there would ultimately be no escape.

Our ancestors probably eliminated the mammoths, along with other preys. Humans changed the species composition of islands, mainland areas, even continents. The demise of mammoths and other herbivores almost certainly changed the vegetational structure of land areas. The emergence of agricultural societies ten thousand years ago heralded the far-reaching impacts of land clearing and water control. Forests began to disappear and natural aquatic systems were modified. Human impacts on natural habitats began slowly; they are now increasing at an appalling pace.

Consider these facts. Primarily as the consequence of selective killing and habitat destruction, more than five hundred species or subspecies of native North American plants and animals have gone extinct since Jamestown was settled in 1607. That's more than one per year! And that's only for this continent. More human-caused extinctions have occurred in this century than in any other, and the incidence is increasing. The species extinction rate that looms for the modern world is enough to scare anyone. The U.S. Fish and Wildlife Service (USFWS) now recognizes more than six hundred species in this country as endangered or threatened. In other words, we are in danger of losing more native species in fifty years than we did in five hundred. Is this any way to run a country, or a world?

Each year the USFWS adds about fifty species to the endangered or threatened species list. However, the figures do not accurately reflect the true numbers that warrant such designation. More than four thousand United States species are candidates for endangered or threatened status. They have not been placed on the list, but this does not mean they are not headed toward an untimely extinction. Their continued absence from the official list is a reflection of how political rhetoric can override proper action. The recognition of species as endangered can hamper commercial interests such as timber operations, mining enterprises, land development. When these interests use financial resources to influence politicians, they often succeed in retarding or permanently blocking efforts to declare certain species as officially endangered.

We know we are causing an environmental crisis. We see the changing landscape and the impending loss of species. Detecting the cause is one step toward a satisfactory solution. And human activities are clearly a common denominator in many environmental changes of the present and past. By recognizing how particular actions affect the world's natural environments and species, we can better propose appropriate solutions.

Some insights can be gained from the findings and reports of the Species Survival Commission of the International Union for the Conservation of Nature, better known as IUCN—the World Conservation Union. This international network of more than one hundred cooperating organizations and about four thousand members provides scientific leadership and counsel for conservation efforts directed toward specific groups of plants and animals. The commission focuses on species in serious straits as well as those potentially in trouble. Biological field-workers constitute the commission, which includes professional ecologists, geneticists, museum curators, and staff members of zoos, aquariums, botanical gardens, and wildlife agencies. In the commission's newsletter, *Species*, reports by specialist groups provide information on the status of various plants, animals, and habitats, and give pause to any clear-thinking person. In some instances the environmental culprits are obvious; in others, the evidence is less clear. In every instance a species is in jeopardy.

One report tells of a spectacular carnivorous pitcher plant from Mount Kinabalu National Park in Borneo, a region with the greatest concentration of pitcher plant species in the world. One of these, the rajah pitcher plant, is enormous, with tubular flasks large enough for a human fist to fit through the opening. These animal-eating plants each hold more than a quart of digestive fluid. An ecologist found two drowned rats inside one plant. These rajah pitcher plants are disappearing at an alarming rate. Even within the confines of the national park, where guards patrol, poaching is relentless. One site in the park was once known as the "rajah kingdom" because of the huge population of the giant pitcher plants. None remain.

The poaching and sale of rajah pitcher plants is attributed to Australian collectors, and the biggest markets are believed to be the United States and Japan. A healthy rajah pitcher plant may sell for as much as one thousand dollars. Anyone who can buy a plant at that price can afford to pay a few thousand dollars toward a fine. But few fines are levied. Who is to blame?

The specialist group for cacti and succulent plants gives a disturbing report from Zimbabwe. Local officials predict that most of the succulent-plant habitats in the country will soon be gone unless strict environmental protection laws are upheld. One recently passed, yet unenforced, law prohibits even landowners from removing or selling native plants. The solution to the problem seems obvious: Enforce the law. But enforcing environmental regulations is not always easy. Whom should we blame?

According to another report, at the beginning of the twentieth century an estimated one million rhinoceroses inhabited the continents of Africa and Asia. As recently as two decades ago, about eighty thousand of these impressive land beasts were alive. Today fewer than ten thousand exist. That would be a disturbing number of rhinoceroses to be roaming around in, say, a shopping mall parking lot. But considering the vast geographic region they once occupied, we can safely declare that we are running out of rhinoceroses. A primary reason cited for the decline is a demand for rhinoceros horn as an ingredient in Chinese medicines. And rhinoceroses are not the only species threatened by Chinese medical practices.

China once had thousands of tigers. Today, fewer than one hundred live in two of the major areas of tiger habitat, and only "a handful" are believed to be in two others. Like rhinoceroses, a handful of Bengal tigers may be too many in some situations. It's a pitiful few when we are considering the species itself. According to the specialist group's report, tiger bones, like rhinoceros horn, are used in making Chinese medicines. This craving for the bones is the basic cause of the decline in tigers in China. Officially, the Chinese have protected tigers since the 1970s, but extensive poaching continues. A tiger-breeding farm has been established in one province, but this effort could never fulfill the demand for bones. Tiger bones are also illegally imported from other countries. Reports of sacks of bones being shipped to China indicate a rise in tiger poaching in Nepal and India.

China is not the only country involved in tiger bone traffic. South Korea imported almost two tons of tiger bones from 1985 to 1990. And in a single year two tons reportedly went to a brewery in Taiwan. Why? To make tiger bone wine. As if making a substance that is disputably a medicine were not shaky enough grounds for exterminating a species. An average tiger skeleton weighs somewhere between twenty and forty pounds. So the brewery alone probably depended on more than one hundred tigers a year. And these figures reflect only the legal reports for which records were kept. Illegal poaching measures its success in the records that are not kept.

The IUCN report identifies cultural legacy as a major threat to animal populations. For centuries, the Chinese have based much of their medical system on the use of animal products. To change ideas entrenched by culture is a difficult, perhaps insurmountable, problem. Suppose younger age groups in China were taught that tiger bones are not essential for making effective medicine? At the present rate of decline, tigers would be extinct before today's youngsters are old enough to influence their society. Not enough time remains for this kind of education if tigers are to continue to inhabit the earth. Tigers can be scary animals, true, but attitudes that condone the total absence of them are even scarier. What happens when one culture's fervor to protect wildlife species conflicts with another culture's belief in the efficacy of animal parts? Now that's a true Chinese puzzle.

The world-scale signs of environmental degradations are disturbing; however, the severity of problems cannot always be properly assessed and agreed upon, even by ecologists. In such instances, the environmental culprits are not easily identified. One still-unresolved global issue is a reported worldwide decline in amphibian numbers. Reports on the phenomenon have appeared in magazines and newspapers, including those my neighbor buys in the grocery store checkout line with such captivating headlines as "Aliens Are Stealing Our Frogs." The perception of some herpetologists is that serious declines in amphibian populations have occurred around the globe during the last decade.

Why should we be concerned if a worldwide decline in amphibians is real? One reason: Amphibians are part of our natural environments. They do not deserve to be destroyed any more than other species that inhabit the earth. Such reasoning may not be compelling for some dollars-and-cents people. So consider another reason, one with economic implications: Amphibians are sentinels of costly environmental disaster. Most amphibians require a healthy terrestrial habitat to live in as adults and a healthy aquatic habitat in which to breed and lay eggs. Small size, moist bodies, and permeable skins put amphibians in constant and thorough contact with soil and water, making them sensitive to altered conditions of either. Because of this dual dependence on terrestrial and aquatic habitats, amphibians are highly responsive bioindicators of environmental stress and degradation. Think of amphibians as an environmental early warning system, an alarm that sounds before the problem becomes too severe—or too expensive—to correct.

In the early 1990s numerous accounts were given of disappearing and declining populations of amphibians, but the culprits remain invisible. The beautiful and highly conspicuous golden toad, a species restricted to a small region in the Monteverde Cloud Forest Reserve in Costa Rica, went unseen in the wild for more than two years. The apparent decline was investigated by Martha L. Crump, Frank R. Hensley, and Kenneth L. Clark of the University of Florida. Golden toads were observed breeding every year for more than two decades, beginning in the 1970s. The investigators found more than fifteen hundred adults in 1987. From 1988 to 1990 they saw only eleven individuals. After analyzing weather patterns, the investigators concluded that warmer water temperatures and drier than normal climatic conditions may have produced adverse breeding conditions. One possibility for the observed decline was a normal population response to an unpredictable environment. Perhaps the toads simply retired to hidden, underground sanctuaries because of an extended dry period. However, other frog species characteristic of the pristine habitat also appeared to be less common than normal, signaling concern that heretofore unidentified environmental factors could be responsible.

Even some populations of the family of common frogs that includes bullfrogs, leopard frogs, and the so-called edible frog of Europe appear to be in trouble. David

B. Wake of the University of California at Berkeley reported that some frogs native to southern California have disappeared or declined dramatically in many localities. The yellow-legged frog, once a common inhabitant of Yosemite National Park, has become rare or absent at many sites. Emphasizing the concern about a global decline in amphibian populations, Wake also noted that several frog species from Queensland, Australia, are thought to be extinct. Among these is the gastric brooding frog, in which the male actually sucks up the eggs and holds them in his mouth until they hatch.

A study cited in *Encyclopaedia Britannica* and published in the scientific journal *Science* was a twelve-year survey of amphibians in a protected wetland on the U.S. Department of Energy's Savannah River Site. Researchers Joe Pechmann, David Scott, and their associates at the University of Georgia's Savannah River Ecology Laboratory captured, measured, and released more than a quarter of a million frogs and salamanders during the study. Daily records of amphibian populations were compiled for more than a decade in a natural habitat seemingly unaffected by environmental degradation. The study documented that natural population levels of some species can fluctuate greatly from year to year, varying from no reproduction in some years to increases of more than fifty thousand larvae in others.

The revelation of the magnitude of natural variability in amphibian numbers gives a clear message that some amphibian species must deal constantly with natural declines. When, in addition to the natural population fluctuations, an amphibian community is subjected to human-caused habitat degradation or to the removal of individual animals on a large scale, some species will eventually be unable to recover. If human disturbance occurs at a time when the numbers of an amphibian population have already ebbed, the impact may be devastating. As demonstrated by the Savannah River Ecology Laboratory project, long-term research will be necessary to determine the disposition of many of the world's species that appear to be in trouble. Do we all have a responsibility to ensure that the research funding necessary to support such studies be provided? Is someone who opposes funding basic research on biodiversity an environmental culprit of sorts?

The phenomenon of a global decline in amphibians does not go uncontested, but for some species, the decline in numbers of specific populations is unchallenged. The disappearance of amphibians locally can often be attributed to the destruction of wetland habitats, vital systems for reproduction of many amphibian species.

Replacing a wetland with a shopping mall and an asphalt parking lot is a common cause of local elimination of native species. In such instances, the culprits are obvious. However, what causes the disappearance of amphibians in seemingly pristine or relatively undisturbed habitats? Although not yet confirmed, some suggested causes of amphibian decline in such habitats are global warming, acid rain, pesticide contamination, and increased ultraviolet radiation at high elevations due to thinning of the ozone layer. If any of these explanations is true, at whom do we point an accusing finger?

103

Amphibians are simply one group that has received focus and attention, but in many instances, local or widespread extinction of many taxonomic groups is not just suspected, it is confirmed. This is particularly evident in tropical regions, but temperate-zone species are not immune from environmental abuse. The status of California's native fish fauna, determined by Peter B. Moyle and Jack E. Williams of the University of California, Davis, is a dramatic example. California fish constitute a taxonomic group from a particular region for which solid background data have been collected. Based on the 1991 report of Moyle and Williams, only 31 percent of California's 115 native fish species were judged to be ecologically secure. A greater proportion (40 percent) were officially listed as endangered (12 percent), threatened (10 percent), or expected to warrant such listing soon (18 percent). Seven percent of the native fish taxa are now extinct. The numerous water diversion projects in the region and the introduction of non-native species of fish are held responsible for most of the observed declines.

Who are the culprits of environmental abuse? We all are in a way. Those who are actively involved in destroying habitats and causing the decline in numbers of a species are blatant offenders. But those who, through apathy, inattention, or timidity, condone such activities are also at fault. Our ancestors may have killed the last mammoth. That does not give their environmentally aware descendants license to continue on the same path.

Most People's Attitude: We Have Met the Environmental Enemy, and He Is Them, Not Us

"Environmentalists complain and preach too much. They think everything we do causes an environmental problem." I feel certain this is the prevailing attitude in some quarters. But is it really what most people think? I decided to take a poll to find out. Admittedly, my poll was small and biased: seven high school seniors recognized as leaders and scholars in their class. Nonetheless the results were intriguing.

All of them felt that we currently face serious environmental problems and that we cannot overemphasize the need for environmental concern. Although they differed on what to designate as the single most serious environmental issue, much as adults do, they agreed on the usual list of pressing problems: global warming, acid rain, industrial pollution, tropical rain forest destruction, disposal of nuclear and solid waste, and everyday littering. And they had no difficulty listing a variety of environmental culprits. The seeming insensitivity of commercial interests to the loss of regional habitats and wildlife was of major concern. Industry, land developers, military programs, and timber companies were not spared. Nor were politicians. The opinion that businesses should absorb the cost of their environmental impacts prevailed. And most of the students felt that consumers should do more for the cause of environmental protection.

All professed a willingness to support laws and to influence political elections in an effort to bring about needed changes. In considering whether any particular approach by environmentalists is ineffective in getting people to be aware of and concerned about their environment, other culprits emerged: vandalism and violence. One student said Greenpeace was committed to a good cause but had tactics too radical to be effective. Another disagreed, declaring that strong actions were the only way to bring attention to some issues. The common thread that ran through their answers was that our environment is being degraded and that appropriate changes should be made. Young people are the world's future policymakers. Those they identify as the culprits of environmental abuse should take heed. Based on my small poll, this includes all of us.

A week after I took this poll, I saw the results of a poll conducted by the South Carolina Department of Health and Environmental Control (DHEC). The results of this larger poll were both gratifying and alarming. DHEC distributed a fact sheet based on the poll results and summarizing its interpretation of the views of South Carolina's citizens. I liked some of the views. Environmental culprits, as perceived by the public, began to emerge.

An overriding public opinion was that we should have laws that encourage people to recycle. Many states already have recycling laws, but someone who prefers to throw beer cans along a roadside rather than return them to a recycling center is seldom penalized in South Carolina. A majority of the DHEC respondents felt that the heroic volunteer recycling efforts being made in many regions were a constructive step but that these programs should be strengthened with recycling laws. The views of South Carolinians on this issue are probably similar to those of people in some other states, many of whom may be unaware that Oregon, Michigan, and Massachusetts, as well as other states, have strong bottle bills. Laws can change attitudes. If the law is a good one, and is enforced, public spirit will soon support the effort.

One startling conclusion of DHEC's poll was that the group people trusted the most on environmental issues was the medical profession. Nurses and doctors received the highest rating. I had assumed that scientists in general and ecologists in particular would have been the most credible groups. I learned the explanation for this apparent anomaly when I called the DHEC unit that organized the survey: Neither ecologists nor scientists were listed as choices on the survey. Imagine taking a poll to determine the best group to diagnose pneumonia and not listing nurses and doctors as a choice.

DHEC meant well with their survey and certainly did a public service by revealing opinions about environmental issues, but declaring the medical profession as the most credible source for environmental issues is deleterious. Enough misinformation about environmental matters is already communicated to the public without adding to the problem by promoting unwarranted trust in a group not trained in ecology. The métier of doctors and nurses is health care, not the environment or ecology. As Rachel Carson said thirty years ago in *Silent Spring*, "We train ecologists . . . but seldom take their advice." Isn't it time we did? Fortunately, in a DHEC poll the following year, a "scientist or ecologist" category was included as a choice. It was rated the highest by those polled.

After the medical profession, the next highest rating went to citizen action groups. I interpret this to include environmentalist organizations, which might number some ecologists among their members. Industry leaders received the lowest rating by the public for providing proper guidance on environmental matters. Elected officials received the second lowest score. My bet is that these two answers would be similar in other states. Industry officials and elected representatives should take note.

In response to a question about the trade-off between economics and the environment, the majority polled indicated that economics, which in South Carolina means manufacturing, agriculture, and tourism, should give consideration to environmental issues. However, as with the nurses-doctors answer on the credibility issue, the DHEC fact sheet sent a wrong message on the economics-environment topic. DHEC stated that its poll provided "strong evidence that the public is not willing to protect the environment 'at all costs' but that environmental protection should be a consideration in economic development plans." What if DHEC had turned the question around and asked, "Would you agree that economic development should be pursued regardless of the environmental cost?" My guess is that the majority of people would have answered no.

A final point from the DHEC fact sheet relates to an issue consistently identified as a primary environmental problem in the United States. Those polled held a conviction that we need a statewide land development plan. However, more than nine out of ten South Carolinians believed, mistakenly, that all natural habitats and wildlife are protected in some way by the state. Only 7 percent of those polled knew that no state agency in South Carolina is responsible for land use planning. Although people nationwide assume that various state agencies control land use, this is seldom so.

Except for game animals and a few nongame species, including most birds, no state or federal protection is afforded the majority of native species on private property in any state. Most nongame species of wildlife have no legal protection. Even federally endangered species of plants have no legal protection on private lands in most instances. If you own the property, you own the wildlife, except game species. No one can legally kill a deer outside the hunting season, no matter whose land it is on; but most states have absolutely no restrictions on what can be done to the majority of our native wildlife. The owner of private land in most places has a legal right willfully to destroy any plant or habitat, and most animals. The private landowner can drain a natural freshwater wetland, cut down an old-growth oak and hickory forest, or pick a federally endangered pitcher plant without breaking any laws.

107

To some people this license to destroy is viewed as an unalienable right, conferred on the property holder through the act of ownership. Yet people accept government's prerogative to acquire private property for the construction of highways; they submit to zoning laws; and in some communities property owners even abide by regulations governing the appearance of buildings. Why, then, should the plants, animals, and habitats that make up our natural heritage not be subject to protective regulation—even on private property? Of course, a responsible private landowner can protect and preserve lands that might actually be more vulnerable if they belonged to the government. Uncontrolled access to public lands is a sure route to habitat abuse. And those in political power often promote projects not in the best long-term interest of natural habitats.

DHEC and other agencies should be applauded for efforts to solicit public opinion on environmental issues. Overall, the findings send a strong message: Environmental matters are foremost on the minds of many people. Some politicians are beginning to realize that many of their constituency have strong attitudes about environmental protection. These constituents may not have the most money; they do have the most votes. Once this relationship is established in the voting booths, stronger environmental laws will be enacted.

Despite increased public awareness of environmental problems, environmental activists can still arouse negative feelings in people. Environmentalists with an emotional stake in an issue often are no better informed about the ecological facts than the group they identify as an environmental culprit. And some exhibit a self-righteous indignation that I call "environmental chauvinism." The word *chauvinism*, used *ad nauseam* in the past decade or so, originally meant extreme, even exaggerated or irrational, enthusiasm for the national glory and military might of one's country. Nicolas Chauvin felt this way about Napoleon's France, even when it was no longer Napoleon's. Chauvinism in the general sense refers to blind support for an opinion or position.

I noticed an example of environmental chauvinism in myself recently while visiting a state game reserve. A group of turkey hunters were roaming the woods, making strange sounds and shooting any unwary gobbler they could lure into the open. I wouldn't have approached such sounds myself, but they apparently appealed to the turkeys. Regardless of one's opinion of turkey hunting, that was not what bothered me. The problem was that the woods and roadside were littered with aluminum cans, bottles, and paper. I thought this odd, people who loved nature and wanted to be in the woods leaving the area in such a mess. And I was downright annoyed about the lack of consideration.

Then something occurred to me: The people who had littered (and who were not necessarily the turkey hunters) were not being malicious, maybe not even indifferent. They were ignorant. They simply didn't know any better. I was being chauvinistic to assume they felt as I did. After all, only in the last few years have the negative aspects of littering natural habitats been emphasized. And a consensus that littering should be eliminated has been especially slow in coming to some places.

We assume a chauvinistic attitude when we conclude that everyone else should be concerned with the same environmental abuses (and solutions) that we identify as important. Perhaps in time we will all recognize the same environmental issues as important. For now, we need a more tolerant approach to situations in which someone has not been educated about what is environmentally proper. This is particularly applicable to actions of the past, which are subject to intense environmental chauvinism. Because of the approaching extinction of the African elephant, some people object to someone else possessing a magnificent piece of carved ivory. Yet not long ago ivory could be legally obtained anywhere in the world. Buying ivory carvings or an elephant tusk was both legal and socially acceptable. Assuming a holier-than-thou environmental attitude toward an ivory owner without finding out the background of the piece in question is chauvinistic.

Such chauvinism is often seen with regard to fur products. Why should someone wearing a fur inherited from her grandmother suffer abuse from an animal rights or environmental activist? Why shouldn't someone enjoy a mink coat that came from a legal mink ranch or was obtained when the sale of wild mink furs was legal everywhere? I agree we need some limits and controls on the acquisition of furs from wild animals caught today. But to object categorically from an environmental standpoint to someone wearing a fur, without knowing the fur's origin, is chauvinistic.

Too often, we get swept up in purposes and missions of the present with little thought to the situations of the past. John James Audubon probably killed more songbirds than some people will see in a lifetime. Shall we judge him retroactively as a nineteenth-century force of environmental destruction? Of course not. The situation was different then. Should we insist that wild African game heads such as those of rhinoceroses and hartebeests be removed from the walls of lodges because we no longer condone hunting of these endangered species? It was legal and socially sanctioned in an earlier time. One might even argue that at that time enough of the big animals were around to sustain the loss by natural replacement. Let us not make *ex post facto* judgments against the big-game hunters of the past. To decry a rhinoceros head that was placed on a trophy wall in the 1950s, when most of us did not recognize the threat to African wildlife, would be unjust.

One class of environmental offenders deserves particular attention. These are the individuals who have the talents, training, and resources to educate others about environmental issues, but who do not do so. Concern has been expressed that university biologists fall within this category because they make little effort to educate the general public about the environment. The problem stems in part from scientists' attitude that they should operate outside the public arena—an attitude due partly to a conviction that the general public cannot comprehend, and should not meddle in, their research.

Many research ecologists do little to inform students on the primary and secondary education levels about the principles of ecology or the importance of understanding natural history. How can ecologists expect the general public to appreciate the intricacies of plants, animals, and natural habitats if they themselves do not get involved in the education process? This trend is changing with the development of environmental outreach and education programs around the country, but we are far from reaching an optimal level of communication.

Paul D. Haemig of the University of Umea in Sweden published an article on a related topic in the *Bulletin* of the Ecological Society of America, the largest group of professional ecologists in the world. Haemig feels that the education system in the United States operates too much from a humanistic attitude and philosophy. Humanism focuses on the capacities, achievements, and values of people, with minimal attention to the natural world. Thus, in the consideration of species and ecosystems, emphasis is placed on those regarded as resources. In other words, our entire education system stresses human interests and gives minimal attention to other species of animals and plants—unless they offer us a direct benefit. Haemig says in his essay that the United States has an educational system that produces "ecological illiterates" who are "unable to identify the plants and animals in their own backyard." This may be a bit of an overstatement, but his message is clear. The appreciation of natural history is seldom foremost among educated people in the United States.

Haemig jabs even harder at university biology departments that do not value a knowledge of basic natural history. As one familiar with botany, zoology, and biology departments at many universities, I must agree. The importance of teaching courses in ecology and natural history is barely recognized. Haemig states that most United States colleges do not even require their biology majors to have a course in basic natural history. Can we expect those incipient biology professors to instill a proper attitude in their students when they themselves often have little or no environmental awareness?

One reason for the lack of emphasis on ecology and natural history in many colleges is that much of the research done by the professors is in cellular and molecular biology. This focus is a result of the high funding levels for anything that can be construed as medical research. Biology departments naturally tend to seek professors who can bring in the most money with their research. Course offerings, in turn, are determined by the specialty areas of the faculty. The result is that biology departments are often overstaffed with cellular and molecular biologists who are not trained to teach ecology or natural history, even if they are interested in doing so.

The time has come for United States colleges and universities to develop twenty-first-century attitudes about environmental issues. An appreciation of natural systems must be encouraged through official recognition of appropriate courses in natural history and environmental awareness. Ideally, universities should require some minimal level of environmental literacy for graduation. Until the attitudes that prevail in many science departments change, universities are going to have a hard time educating students about the importance of natural environments.

Haemig suggests a corollary for why biology courses that provide exposure to natural history through field trips are limited in United States schools from the primary to the college level: our acute dread of lawsuits. To explore nature, people must go outdoors, and they can indeed get hurt out there. A teacher who keeps the class in the classroom avoids lawsuits involving snakebite, insect stings, and poison ivy. This is a terrible indictment of the United States legal profession and a deplorable comment on the litigious nature of the general public. But "litigaphobia" aside, the basic problem remains. Most education professionals neither demonstrate nor teach the proper degree of appreciation for nature.

There is yet another, related group of environmental culprits: people who are environment-minded and supportive of environmental protection programs but who do not always do their part with logic and thoroughness. Nat Frazer of Mercer University suggests that sea turtle enthusiasts, a group that has been highly publicized in the environmentalist movement for many years, represent this sort of right-minded, but misguided, environmentalism. Frazer gave the keynote address at the Tenth Annual Workshop on Sea Turtle Conservation and Biology, held at Hilton Head, South Carolina, in 1990. His primary message was that we have become enamored of partial solutions, or halfway technology, in addressing environmental problems. Halfway technology is a term used in a medical context by Lewis Thomas, though Frazer applied it to sea turtles. His points are pertinent to attitudes about environmental management and mismanagement throughout today's world.

Sea turtle conservationists are involved in protecting nesting turtles and eggs around the world. Thousands of eggs are incubated each year in artificial hatcheries; later the hatchlings are released into the ocean. In some programs, the babies are reared in laboratories for months; they are not released until they reach larger, "dinner plate" sizes. These so-called head-start programs presumably assure that young turtles are spared from predators and other hazards that a silver-dollar-size baby turtle would face at sea. Unquestionably, these programs have released millions of baby turtles that would have perished before they got beyond the breakers.

Such programs, as Frazer points out, make good press. News programs with volunteers releasing baby turtles into the surf are popular. But he also identifies a major weakness in such conservation efforts. They save baby turtles at the beach without solving the problem of why every species of sea turtle in the world today is threatened or endangered. Some environmentalists promote an impression that increasing the number of hatchling-release programs is a solution for dealing with the abuses that cause the problem; that is, if the sore gets worse, apply a bigger bandage.

Frazer specifically criticizes the head-start programs. In the first place, no one knows if the juvenile sea turtles raised in captivity are any better able to survive than those that enter the ocean as babies. No head-start individual has ever been confirmed to reach maturity in the wild. Second, no consideration is given to how the rest of the ocean's natural community of organisms is affected by the removal of these individuals from circulation for an extended period of time. In other words, young turtles presumably play some role in the ocean's environment. If we prevent their presence, does the ocean ecosystem function normally? Are we creating another environmental welfare program?

The root of the sea turtles' problem is that they are threatened with habitat destruction and overexploitation. Two major, yet controllable, human-caused situations confront sea turtles. One is the incidental taking of adult sea turtles by shrimp trawlers. The other is the continued increase in beachfront lighting along the sea turtles' ancestral nesting beaches, a factor that can preclude nesting by females and disorient hatchlings. The fact that both impacts can be dealt with successfully through appropriate technology is both encouraging and disheartening—encouraging because the solution is within our grasp, disheartening because we have not applied the solution on a wholesale basis.

The use of TEDs (turtle excluder devices or trawling efficiency devices) greatly reduces the direct mortality of turtles in offshore areas. Unfortunately, but understandably, although TEDs are required by various state and federal regulations, the forced use of an expensive attachment for a shrimp trawl is not readily accepted by some members of the shrimping industry. The final resolution of the killing of sea turtles by shrimpers will reside in economic considerations. Maybe this is an instance when (as the students I polled believed) consumers should do more for the cause of environmental protection. Perhaps we should prepare to pay more for a shrimp dinner to enable every shrimper to buy a TED and use it properly. As for artificial beach lighting, the answers are simple. Lights can be designed with deflector panels or placed closer to the ground. Low-pressure sodium vapor lights seem to have a less negative effect on nesting sea turtles than do mercury vapor lights. A more radical approach, which I support, is to prohibit lighting on an open beach.

Halfway technology may help us release more turtles each year, but it does not solve the problem. Sea turtles continue to face the same problems their parents could not endure. We need to fulfill our environmental responsibilities on a deeper level by seeking permanent prevention of the problems rather than temporary cures. Sea turtles are the example used by Nat Frazer, but endless examples could be made of other animals, plants, and natural habitats throughout the world. Halfway technology is not a solution. It is merely an attitude, a way to rationalize our irresponsibility in attending to the real problems confronting the world's environments.

The environments of the world are under siege, and the assailants are numerous. All of us are to some extent guilty of environmental abuse, though the spectrum of abuse runs from pitch black to endless shades of gray. Paradox and irony lie in the fact that each of us also qualifies as a victim of that abuse.

113

The Riddle of Reelfoot Lake

On some environmental issues the question of who is right and who is wrong presents a quandary. Even when we take a careful, objective look, some issues defy easy resolution, especially when both sides present plausible arguments. This is the case with Reelfoot Lake, one of the more unusual aquatic habitats in the southeastern United States. Located in Tennessee, Reelfoot Lake actually presents two riddles, an ancient one and a modern one.

The ancient riddle concerns the origin of the lake. Earthquakes in 1811 and 1812 are believed to have opened up the ground within a huge cypress forest, leaving a fifteen-mile open fissure with its western tip alongside the Mississippi River. After 180 years, truth and conjecture are a bit hard to unravel. But some say the Mississippi River flowed backward as it filled thousands of acres of sunken ground. Chickasaw Indians believed the ground sank and wiped out the tribe of Chief Reelfoot. Angry medicine men of a tribe to the south had predicted something like that, since the chief and his braves had treated the other tribe unfairly. Whether the cause was shifting tectonic plates, Indian politics, or both, Reelfoot Lake stands today.

The modern riddle of Reelfoot Lake is how to manage this natural wonder fairly. The lake teems with gar, bowfin, and all species of game fish found in the region. More species of freshwater turtles live there than in any other lake I know of in North America. Bald eagles flock to the area by the dozens in the winter, and hundreds of waterfowl and wading birds are present year-round. Whether you like to fish, hunt, bird-watch, or just float in a canoe through a majestic forest, you can find outdoor recreation you enjoy at Reelfoot Lake.

Despite uncontrolled timbering in the region during the past century, massive cypress trees dominate a vast expanse of water, forming swamps big enough to get lost in. During times of low water levels, an outboard motor will not get far before it hits a hidden stump, even in what looks like open water. The last time I went, the water was three feet higher than the time before, when thousands of cypress stumps could be seen, and felt. This changing of the water level, controlled by a spillway built in the 1930s, has led to a hot controversial issue in the region, one relevant to many environmental problems.

The Tennessee Wildlife Resources Agency (TWRA), a state agency, contends the lake is undergoing natural succession. This means the lake is filling in with organic sediments because of heavy shoreline vegetation. So the TWRA has developed a plan to raise and lower the water level to keep aquatic shoreline vegetation in check. The process is used for lakes and reservoirs in many areas. The TWRA justification is that the fish would thrive, thus maintaining the lake for recreation. Plans include buying several thousand acres of farmland around the perimeter to increase the size of the wildlife sanctuary. To the conservationist this sounds reasonable—enhancing a habitat for a variety of uses. But you don't have to talk to many people who live there to hear some other sides of the story.

If the state acquires the surrounding land, what are the people who run farm-related businesses to do for a living? There's no more farmland to buy in the area. The people who own feed stores or sell farm equipment foresee a major decrease in the farm market in the region. The people whose livelihood depends on renting fishing or sight-seeing boats also have a problem. Drawing the lake down leaves their docks high and dry for weeks or months. I talked with one owner who had to close during part of the summer because of a major drawdown. Most claim they will go out of business if forced to shut down for an entire season. Motel resort owners around the lake likewise envision a loss in profits. The majority of motels could be eliminated if the lake is not attracting tourists. Besides the reduced water level, the smell of rotting organic matter also discourages tourism. The TWRA takes the position that it is in the long-term best interests of such businesses to endure the short-term disruptions. It contends that occasional drawdowns will maintain the viability and productivity of the lake the businesses depend on.

Reelfoot Lake presents a situation in which many small enterprises could be terminally affected by what seems best for environmental interests. Further complicating the issue is the claim by some people that evidence shows the state's plan will not accomplish what it is intended to. Then there's the environmental philosophy that maintains the "natural" thing to do is to leave the lake alone. Don't manipulate it at all; let it proceed with natural succession. The TWRA holds that its plan will reintroduce lake level fluctuations that are similar to those before the spillway was constructed. So, what is the "natural" thing to do?

The answers for how best to handle a situation like this are no easier than deciding how to deal with a multitude of other environmental stalemates in which people have different interests. Such problems are going to become more and more common around the country and the world. The modern riddle of Reelfoot Lake is the environmental riddle of our time.

A comparable dilemma that most North Americans are familiar with involves the spotted owl and the ancient forests of the Pacific Northwest. How many trees have been cut down to supply the paper to publish everything that's been written about the spotted owl? This includes reports by the U.S. Fish and Wildlife Service aimed at protecting the owl, legal documents submitted by timber industries to protest the action, and tons of newspaper and magazine articles. This bird has gotten more attention in the early 1990s than most people do in a lifetime. And which side you take in the controversy may depend on whose story you read last.

Cogent arguments can be presented from different sides. One side maintains that the bird is a symbol for the preservation of the old-growth forests of the Northwest. The enemy is a timber industry that can't bear to see a big tree standing upright. The defenders of the owl take the position that we should protect the bird to save the forests. Another side asserts that thirty thousand lumber industry families are more important. The enemy is an army of eco-shriekers who wouldn't know a board foot from a knothole. This side says the forests are a natural and renewable resource that serve not only the region but the entire country.

The issue has become one of emotion and politics, not one based on economic and environmental logic. Is there an unequivocal right and wrong side? The owl is a species with a right to exist (and is perhaps an ecological barometer of pressures on the environment); it is also a legal symbol that can prevent cutting of the forests, the basic objective of environmentalists. Having free access and timber rights are the objectives of the lumber industry.

Environmentalists maintain that the old-growth forests are an environmental legacy that belongs to all of us. The forests induce awe in all who enter them. They stand as a hallmark of our natural heritage. We have no right, no need, to destroy them. We should live with them, intact, as humans have done for the last several centuries. And what about those thirty thousand families? Well, we might ask what those lumber industry families have planned for a few generations from now, after they have completely eliminated the old-growth timber. There will probably be twice as many of them by then. What kinds of jobs do they have in mind to pursue when the forests are gone? Could they begin those occupations now?

But look at the matter from the perspective of a tree cutter whose parents and grandparents were tree cutters. Suppose you had grown up in these forests and loved them just as much as somebody from the city who visited them once a year. You know the trails, the trees, the animals, and you know the wealth of forest left to be harvested. You also know that what has been cut will grow back. If the timbering stopped cold, your family would have no income. Would these people who want to save the owl and the big trees be willing to pay your salary for a year or so while you found another job? You don't think so.

You note that most of the people who rant and rave and write about your insensitivity to the forests don't even live near them. Many have never even seen these forests. Why are they more concerned about an owl than about you and your family? You would bet that a good portion of the people complaining about your cutting down trees grew up in houses made from lumber that came from your forests. No one complained then.

Logical arguments will not resolve the issue. Many hard feelings are yet to come; and new environmental threats ensure that the battle between environmentalists and business enterprises will not abate. For example, the USFWS proposal in 1991 to make the marbled murrelet, a seabird that nests in coastal forests of the Northwest, a threatened species will strengthen the position of environmentalists. Listing the bird will result in protecting many coastal forests from timbering. These forests, beyond the range of the spotted owl, do not benefit from its protected status. We may expect a lot of legal battles, and a lot more trees to be cut for paper, before the full issue is resolved. But we should remember this: If the owl goes extinct, even when the trees return, we will not have the same forests we have now. The owl, as much as the trees, makes these forests what they are.

The dilemmas presented by Reelfoot Lake and the spotted owl are regional, but

the problem they represent is universal. Those unaffected directly by the socioeco-
nomic impact of a situation can readily claim that a habitat and its species are too
important to lose. An individual's choice of sides is influenced not only by the
information available on short-term and long-term environmental consequences
but by the immediacy of a decision's impact on him or her.

In considering whether to manipulate a habitat, as has been done at Reelfoot
Lake, or to let it take its natural course, we always risk making the wrong decision
—and having to live with it. One type of environmental action that can have dire
consequences is the purposeful introduction of alien species to a region. In an effort
to solve one environmental problem, we often create a more serious one. In addition,
we will always have to live with accidental introductions, species that arrive from
foreign lands by chance.

Many species of plants and animals introduced into the United States from
other countries have become problems—kudzu, fire ants, Japanese beetles, starlings.
Some of these unwelcome species, such as fire ants, which made landfall in Mobile,
Alabama, a few decades ago, arrived on our shores by accident. Others, including
the starling, were intentionally introduced into the United States for a variety of
reasons, many of them ill conceived.

In the 1890s, a pharmaceutical manufacturer named Eugene Scheifflin lived in
New York. He admired two things—Shakespeare and birds. Scheifflin's fondness
for Shakespeare's works led him to undertake the ambitious task of importing to
this country all the birds mentioned in Shakespeare's writings. Through his repeated
efforts, starlings were successfully established in the United States. As early as 1895,
people were expending time, money, and energy to eradicate the descendants of
Scheifflin's first few birds, and the efforts continue today.

Ours was not the first country to regret the introduction of the starling. Starlings
were introduced into New Zealand in 1867 and were becoming a problem before
1880. Despite New Zealand's negative experience, some United States bird lovers,
along with Scheifflin, felt these birds would be an economic boon because they ate
insects.

They will eat insects. But much of their diet is fruit and grain intended for
human consumption. A starling does not "eat like a bird"; individuals consume one
to two times their own weight each day. They cost ranchers millions of dollars a
year in beef feeding lots by eating the cattle's grain or contaminating it with their

wastes. The commotion alone from a flock of starlings, who often travel in large noisy groups, drives many native songbirds from their nesting sites. The enormous flocks also annoy people, and the Civil Aeronautics Board is troubled by such flocks when they congregate on or near airstrips because of their potential for causing airplane crashes.

All efforts to control or eradicate the starling by biological means have been thwarted by the species' amazing reproductive capacity. In 1965, twenty-five million starlings were estimated to be in Virginia's Dismal Swamp and more than a half billion were believed to be in the United States! Starlings are native to Europe. Breeding from the Scandinavian region and Siberia to as far south as India, Africa, and Spain, they do nothing in moderation. They are rapid reproducers, voracious eaters, and noisy encroachers on the nesting sites of native birds.

Non-native species continue to be intentionally introduced in North America and other parts of the world. The reasons are varied and often controversial; some, such as the one that caused the starling folly, are whimsical. Sometimes a food source or game species on one continent has promise for another. Ring-necked pheasants and honeybees are both recognized as successful immigrants to North America. Kudzu was brought from the Orient to establish a fast-growing ground cover for eroded areas. Sure enough, in some parts of the kudzu belt you can't see erosion—or anything else, except kudzu. Sometimes a species is introduced as a biological control for a pest species, often one that was itself introduced. Many introductions, including Norway rats and fire ants, are accidental.

Ecologists differ in their opinions on whether the practice of deliberately introducing non-native species is environmentally sound. A safe statement would be that some introductions have worked to our advantage, and some have not. Another, equally safe, is that we cannot predict the outcome. Whatever justifications are offered, the long list of problem species bears a message. We need more thoughtful consideration of the potential ecological consequences before we attempt to modify natural systems. Nonetheless, species introductions will continue, at least accidentally, and sometimes with drastic results.

With the possible exceptions of Hawaii and southern Florida, where escaped imports from other regions are becoming commonplace, one might ask why we don't have more problems than we do with exotic species. Birds, snakes, and other pets from foreign lands escape from their owners, yet they do not take over like the starling. Some species do become established in localized areas, surviving but not dispersing into other regions. The clawed frog of Africa thrives in an area of California, and tropical gecko lizards can be found in many port cities in the South. Perhaps environmental conditions encountered by such species are different enough from their native habitats to control their population increases. Or perhaps one day they, too, will become problems. The starling apparently found itself in an unexploited ecological niche with no natural predators, no environmental controls, and no successful avian competitors. Who knows? Maybe starlings are a replacement for the passenger pigeon. Now that ought to teach us something.

The introduction of non-native species can be a world-class problem. Ask New Englanders how they feel about a fungus called *Ceratocystis,* the one that causes Dutch elm disease. Or Southerners about fire ants. Or Australians about rabbits. Two dozen European wild rabbits were brought from England in 1859 and released on a sheep ranch near Geelong, Australia. Their release in the southeastern corner of the country marked the beginning of a war between Australian farmers and the furry intruders.

Rabbits, native to Europe, Asia, Africa, and the Americas, are ecologically important in natural ecosystems throughout much of the world. However, the rabbits taken down under, where rabbits did not naturally occur, caused a problem. The rabbit impact was beginning to be felt before the clipper ship they came on, *Lightning,* had barely cleared port. (The name *Lightning,* as you might guess, is not held in fond memory by Australians and is unlikely to be used for their America's Cup entry.)

As we all know, rabbits produce many young, and they produce them often. Only six years after the two dozen rabbits were released in Australia, twenty thousand were killed in the neighborhood where the forebears had been set free. Probably more than a thousand rabbits a day were being born in Australia by that time. A high reproductive rate is not the only explanation for the rabbit's success in Australia. Rabbits had few natural predators to control their numbers, and the nutritionally poor pastures of the region, unsuited for rabbits, were replaced with high-quality Mediterranean grasses. By the early 1900s they could be found throughout the entire southern half of the country. In only fifty years, the Old World rabbit had spread from one small farm in Australia across an entire continent, a distance of more than two thousand miles.

Australia's native grazing animals, such as kangaroos and wallabies, have suffered from this miniature hopping menace, but rabbits have posed their most serious threat to the native vegetation. According to Ian Parker of Australia's CSIRO Division of Wildlife and Ecology, the full impact of what rabbits have done and are doing to native vegetation is yet to be seen. The Australian sheep industry has felt a major economic effect. Wherever large rabbit populations are found, sheep raising is difficult. Sheep and rabbits prefer the same type of vegetation, and rabbits seem to be more successful at getting to the food than sheep are.

An attempt at dealing with the rabbit menace was made by introducing yet another species as a biological control, a virus called myxomatosis. Rabbits infected with myxomatosis died quickly, and at first the virus appeared to be a solution to the overabundance of rabbits. Indeed the numbers dropped sharply, for a while. But between four and seven years after the virus was introduced, rabbits continued to be a problem in Australia because of a biological phenomenon that no one predicted —coevolution, the evolution of two organisms simultaneously in response to one another.

In Australia, the myxomatosis virus must pass between rabbits (the host) and mosquitoes (the intermediate host) in order to complete its life cycle successfully. The viruses are transmitted from one rabbit to another through mosquito bites. But as a consequence of evolutionary phenomena, both rabbits and viruses changed in response to the new set of circumstances. Today the rabbits in Australia are less affected by the myxomatosis virus than they were originally. For the rabbits, the evolution was straightforward. Some individual rabbits were less susceptible to the virus than were others. Those genetically disposed to resist the virus were more likely to contribute to future generations of rabbits, so virus-resistant genes became common in the rabbit population. Meanwhile, the myxomatosis virus was also playing some evolutionary odds. If the virus killed its furry host before it was bitten by a mosquito, the virus died along with the host. But some strains of the virus let an infected rabbit live longer. These viral strains were more likely to be picked up by a mosquito and thus had a higher probability of being passed on to other rabbits. These less-lethal strains became successful and prevalent throughout the Australian rabbit population. Today, Australian rabbits are more resistant to myxomatosis and the virus is less virulent. Although the intermediate strain of the virus is an effective control to some degree, Australia can look forward to having rabbits as a continuing part of its fauna.

The intricate relationships of plants and animals in a region are delicately balanced. Because we do not fully understand ecology, we make many mistakes in our attempts to manipulate the natural world to our own advantage. The rabbits of Australia serve as a distressing example of such a mistake. Judging from previous experiences with the introduction of new species, Australians may even find that the cure will prove to be not only ineffective but a problem in itself. If the myxomatosis virus evolves strains that can survive in native or domestic animals, will Australians forgive those who thought they were doing the right thing?

A little environmental malice and a lot of environmental ignorance lie within each of us. Our challenge is to accept the environmental mistakes and shortcomings of the past and gain consensus on some major goals for the future. We must agree that environmental education is essential for everyone and that preserving native biodiversity in every region is imperative, even at the expense of possibly slowing economic progress. We must develop an international attitude that curtailing the loss of wildlife species is a foremost priority. Actions that clearly result in environmental abuses must not be tolerated, but we must also evaluate and reflect before we take irrevocable environmental stands. We need to be aware of the depths of our innocence and ignorance about what is environmentally proper.

Add People,
Subtract Wildlife

Choose any environmental problem you consider urgent—ground and surface water pollution, global warming, tropical rain forest destruction, ozone depletion, radioactive waste disposal. Give the problem some careful thought. No matter what environmental issue you consider, the underlying cause is the same: Too many people inhabit the earth. Too many of us are trying to take advantage of limited resources. Too many of us need or think we need the earth's resources. As a result, the resources dwindle daily.

In 1800, fewer than one billion people, about seventeen humans per square mile of land, lived on earth. We are now approaching one hundred humans per square mile. If the oceans rise from a global warming trend, we'll have even less land to live on. A vast portion of land is already under agricultural or urban development. More people will mean more land devoted to crops and more land for housing. Where will this land come from? More human land means less natural land. Less natural land means fewer native plants and animals. The formula is quite simple. Increase the people; decrease the biodiversity. More of us means less of everything else.

In the minds of some people, critical social and economic problems of our times, such as overcrowded prisons, unemployment, and homelessness, may take precedence over environmental issues; but these problems are also the product of overpopulation. According to Zero Population Growth, an organization established in 1968 in response to the world's growing population problem, babies are born at a rate of more than a third of a million per day. Mexico City alone has twenty-two million people. During the next year, the worldwide increase in the human population will equal four more Mexico Cities. We have a global problem, and many seem unwilling to admit it. Consider how few elected officials publicly acknowledge our basic problem of overpopulation, Vice-President Al Gore, who wrote *Earth in the Balance,* being a notable exception. As we continue to ignore or dismiss the population problem, more and more people are around each year to not think about it, or at least to not concede it is a problem.

The question of overpopulation always solves itself in the natural world. Disease, starvation, parasitism—all eventually increase when a species becomes too abundant. Humans are subject to the same natural laws as other animals, as is evident throughout the world for anyone who cares to take a careful look. These natural solutions, with the uniquely human addition of war, will eventually curtail our numbers. But another uniquely human attribute can prevent our becoming the victims of nature's solutions to the problem of overpopulation. We have the ability to plan, to influence our future, to make decisions. A combination of natural forces will unquestionably control the size of the human population, as it does for every other species of animal on earth, unless we establish a survival plan.

Human overpopulation must be discussed as an ecological issue, not an emotional or religious one. To end overpopulation on a planetary scale, we have two alternatives: We stop adding people so fast, or we subtract some of those already here a little faster. We decrease the birthrate or increase the death rate. We have no other choices.

But the first step is to agree that we have a critical problem and that solving that problem will indeed provide a higher quality existence. Until we reach consensus on this, no agreement can be reached on how to control our population levels. Not facing the problem could leave us eventually with a planet half-covered in asphalt and cement, a planet on which several billion people can visit one another, for a while. Surely this is not what we want.

No matter what moral, ethical, or religious justifications are given for uncontrolled human population growth, our planet cannot withstand a population increase indefinitely. Natural laws are not influenced by even the most eloquent justifications.

An overpopulated earth may survive for a limited time, but eventually the maximum population that can inhabit the earth will be attained. When that happens, natural laws will assert themselves to control the population. And they will not be swayed by political rhetoric or emotional arguments or apologies.

A central aim in creating environmental awareness must be to teach everyone that only so many humans can live comfortably on the planet and that we must therefore control human population growth. Our planet already supports all the life it can. Arguments that we can produce a life-support system for an endless number of humans through modern technology and agricultural practices are meaningless. If we double the number of people, we halve what the average person gets from the finite resources available. In addition, we diminish the world's wildlife through inevitable habitat destruction. Eventually the resources will be insufficient to meet the human demands without annihilating all natural habitats. Do you really want to live in a world comprising only people, pavement, and agribusinesses? Do you want to bequeath such an environment to your descendants? What are they to do when no land remains for more fields, more houses, more parking lots? Technology is touted as a solution to myriad problems, but technology cannot protect natural environments from a burgeoning human population.

In the end, technology cannot even protect us from ourselves. Not only are our native wildlife and natural habitats adversely affected by human overpopulation, we, too, suffer the consequences of our uncontrolled population growth. The quality of such basic requirements as air and water has declined in countless regions of the country and the world. The atmosphere and groundwaters are polluted because too many people drive cars and rely on industries that discharge wastes into the air or water. We are destroying our world. And the destruction is a direct result of human overpopulation.

The basic requirements per person for animal parts, trees, and land are far greater than they were in the past. People and transportable goods travel farther than ever before. Plastics that require fossil fuels, paper that requires trees, and irrigation farming that requires water cause an impact on these natural resources never before experienced on earth. The demands per person are enormous, and they increase daily.

Furthermore, compared to fifty years ago, *twice* as many people are now making those demands. If we continue to multiply at the current rate, the world's population will double again before the grandchildren of today's third graders are in high school. Imagine your community with twice as many houses, twice as many people, twice as many cars. Of course, nature or war may well limit the population growth before then.

The Worldwatch Institute is a nonprofit organization created to analyze and focus attention on global problems. According to its magazine, *World Watch,* its goal is "to help reverse the environmental trends that are undermining the human prospect." The problems it identifies as critical—destruction of forests, thinning of the ozone layer, population growth—are indeed worthy of worldwide concern. To reach its goal, the institute endeavors "to raise public awareness of these threats to the point where it will support an effective political response."

The issues of *World Watch* that I have seen have been outstanding, with carefully researched and accurate accounts of environmental problems on a global scale. Not everyone will enjoy such a magazine, not that people mind hearing bad news. Indeed, the news media thrive on our enthusiasm for other people's misfortunes. But the bad news in *World Watch* is too close to home. The misfortunes are the readers' own, and to correct the problems, they will have to change their own life-styles and to support deep changes in policies, attitudes, even their—our—very culture.

A report in *World Watch* (1990) on the use of cars is indicative of its hard-hitting journalism and provides a glimpse of the sort of radical life-style changes required. The title of the report, by Marcia D. Lowe, is "Alternatives to the Automobile: Transport for Livable Cities." The indictment of the overabundance and overuse of cars is difficult to refute. As stated in the report, massive traffic jams, air pollution, and politically risky oil dependence now plague most industrial nations. Four hundred million cars in the world are the single largest source of air pollution. These cause 13 percent of the carbon dioxide emissions from fossil fuels. In the United States, more than 40 percent of petroleum use is for cars and light trucks.

As for traffic jams and congestion, anyone who has driven in Los Angeles or New York City lately knows why they would rather not. In Seoul and Rio de Janeiro the rush hour is now more than twelve hours long—half the day! More than a quarter of a million people throughout the world are killed in traffic accidents each year. In short, we have too many cars for our own good. Yet with a shift in values and attitudes we could rapidly develop a balanced system of public transport, cycling, and walking. Reducing our overdependence on personal vehicles would make metropolitan areas more livable. Human health would be enhanced through more exercise and cleaner air. The route to efficient transportation and a cleaner environment lies in ending the dominance of the car.

One way to lessen the automobile's stranglehold on our society is to provide fast, clean, safe, accessible public transportation. This would mean lower fuel consumption and cleaner air (and maybe even more pleasant commuters). Consider the following efficiency ratings from the Worldwatch report. A one-lane subway system can transport seventy thousand passengers per hour past a single point. A single-lane bus system can move thirty thousand. A double lane of private cars moves fewer than ten thousand—and that's with four passengers per car. Think of the difference in fuel costs per person if we embraced public transportation.

We, as a society, should make a commitment to lessen our dependence on privately owned gasoline-powered vehicles. Perhaps we should push for government-funded research on solar-powered transportation. Maybe a major increase in gasoline taxes would oblige motorists to become more efficient. A five- or ten-cent tax hardly makes anyone reconsider whether to drive alone or carpool; but what if the tax were a dollar a gallon? Designating or redesigning sizable sections of cities as walking or public transportation areas would be a major step.

Trying to think of effective ways to persuade the average American to accept the need to decrease his or her use of cars is discouraging. But we are clearly on a road to disaster if we do not do something soon. Cars are a prime example of excess and waste in our society, of an unhealthy dependency on technological achievement. But tradition, culture, and convenience are powerful forces to overcome, even when logic and good sense are unquestionably on the side of change. Any attempt to change "the American way" must contend with some formidable powers. Would politicians who receive support from the oil industry promote the necessary changes? Would corporations that make enormous profits from our love affair with the automobile (and, in turn, make financial contributions to politicians) support measures to reduce fuel consumption?

Too many people with too few natural resources guarantees a low-quality existence. We can improve the quality of life by retreating from the blind worship of industrial technology, curbing overpopulation, and developing an appreciation of the natural world around us. For elected government officials to support such measures may seem like political suicide. Not to do so will be a form of genocide.

Each of us and all of us are the environmental culprits of the world. Because of our special needs and opportunities as reasoning, thinking organisms, we deal with the world in a way no other species does. We have an obligation to be aware of how our dealings affect all species, including our own.

PART FOUR

Curbing Environmental Destruction

Something can be done about the devastation of tropical forests, worldwide pollution, and uncontrolled commercial development. Aside from the obvious solution of reducing the increase of human population through effective birth control, we can actively control our environmental destiny in a variety of ways. Many, perhaps most, individuals support the preservation of natural systems and would be willing to make personal commitments to assure such preservation. However, people are often unsure about how they can have an influence on a worldwide, regional, or even local scale. What can one person do?

Individual efforts, when directed at the right targets, can have a positive effect on the preservation of natural systems. Actions or crusades to save natural environments have been launched from a variety of positions. Being a thoughtful consumer, developing political responsiveness, and becoming involved in environmental education are some options. Voters might convince those in the political arena that their constituencies want long-range, sustainable approaches to the natural world rather than shortsighted tactics that promise short-term economic gains but guarantee long-term environmental losses.

Individuals can also join environmental groups. The strategies of such organizations, from the near violent to the benign, provide a composite of actions that is highly effective in mitigating some environmental abuses. One of the strongest, most enduring efforts any individual can make is to cultivate among children a respect for and appreciation of the natural world. Environmental education underscores the importance, indeed the necessity, of maintaining the world's biodiversity. An environmentally educated society cannot plead ignorance about whether measures must be taken to assure the welfare of our natural environments and the diversity of species that constitute them. Part of this education should be directed toward a basic understanding of the natural history of different species. Each of us should contribute in some way to the protection of the world's ecosystems—and persuade others to do so, too.

What Can an Individual Do?

I saw an opportunity once to make what I thought would be a modest personal contribution to environmental preservation. The management of a South Carolina coastal resort told me about a large alligator that had been pestering golfers. The management intended to notify the state that the alligator was a nuisance, which meant it could be legally killed. I asked if, instead, some of us from the Savannah River Ecology Laboratory could catch the animal and remove it to another habitat on the island—one from which people were excluded. They agreed.

The nuisance turned out to be a mother alligator. She didn't intend to be a pest, but people kept hitting little white balls down close to the lake where she and her babies lived. Female alligators being notorious for protecting their young, she would emerge from the lake, chase the golfers away, and occasionally eat a golf ball. Catching an alligator should be no problem for trained professionals from an ecology lab, but before the encounter was over, I had some question about how trained we were and whether we should really be classified as professionals. Research ecologists should be reminded occasionally that they do not know everything about animals, plants, and the environment. Alligators have effectively brought this to our attention more than once.

I took two students, Jeff Lovich and Tony Mills, out at night to make the capture. With Tony being six four and Jeff six two, I anticipated no problem in handling an alligator that was only six feet long. Michael, my twelve-year-old son, came along to watch. When we got to the lake, we saw a pair of red eyes as bright as rubies reflecting from our flashlights. The red eyes were surrounded by what seemed like a swarm of fireflies on the water's surface—two dozen pairs of little yellow eyes.

We had a noose attached to a cable on a bamboo pole. When the mother came near shore, we planned to slip the noose over her head and pull it tight around her neck. We would then put big rubber bands on her snout to keep the mouth closed and carry her back to the jeep. That was our plan. Her plan was different: She just swam around in the middle of the lake with her babies. We devised a new plan: Catch one of the babies. Since baby alligators make a yelping sound when in distress, the mother should come over close to shore to investigate. When she got close enough, we would use the noose and that would be it.

Most of the babies were with the mother, but a few adventurous ones were in the vegetation along the shore. We walked around the edge of the lake and caught one of these. It immediately started making the sound of a frightened baby alligator, and to our satisfaction, along came the mother. The two crimson eyes headed straight toward the shore, fast. I handed the baby alligator to Michael; the rest of us hid behind two big pine trees.

As the mother reached the shoreline, Tony got ready to jump down and use the noose. Only she didn't slow down at the water's edge. The next thing we knew, she was up on land and heading toward Michael, who was holding the baby up in the air and saying, "Dad, Dad, what do you want me to do now?" Being trained professionals, we each offered expert advice. Jeff said, "Climb a tree!" Tony said, "Throw the baby in the lake!" I said, "Run!" Michael turned and disappeared into the woods, still holding this squeaking toy of an alligator. A scared twelve-year-old can run a lot faster than an angry alligator, but the mother was still in pursuit.

She seemed to be moving pretty fast when she passed the three of us, but Tony managed to slip the noose over her head, and Jeff and I grabbed the bamboo pole. Suddenly she reversed her direction, catching the three of us completely by surprise. She turned back toward the lake, dove into the water, and plunged to the bottom. Unfortunately we all had good grips on the pole. The three of us were yanked down the slippery bank into the lake.

The noose had slipped off, and with thoughts of being in the water with an irate mother alligator, all three of us managed to scramble out almost as fast as we had gone in. We found Michael, returned the baby to the water, and went home defeated, with some discussion about safer and less humiliating lines of work. The next day we returned with reinforcements, caught mother and babies, and safely removed the family to a marsh on the far side of the island.

Assisting in an alligator relocation program is not my recommendation for what every individual should do to help make a better environment. I myself help turtles cross roads, answer calls from people who have snakes in their yards, and end up with any number of orphaned baby mammals and birds each year. Although saving individual animals in this manner probably does little to preserve a particular species in a region, it does demonstrate to others that our native species are important and not to be taken for granted.

133

Not everyone wants to become personally involved in protecting animals, nor should they have to. There are many other ways to contribute to our overall environmental welfare. The EarthWorks Press published a book called *50 Simple Things You Can Do to Save the Earth*. In one-to-two-page presentations, the authors give the background of an environmental problem, a few facts reinforcing how serious the problem is, and some specific actions you can take to help.

They suggest some simple paths to a better environment. One presentation, "Stop Junk Mail," explains that it takes almost one and a half trees to produce the paper that makes up the mail each of us wishes we had not received in a year. If everyone in the city of Atlanta (2.8 million) stopped receiving junk mail, we could cut down four million fewer trees each year. The junk mail received by Americans during a single week could produce enough energy to heat two million homes during the winter. This includes the 44 percent of such mail that is thrown away without even being opened.

So, what can you do about junk mail? You have several choices. You can write Mail Preference Service, Direct Marketing Association, 11 West Forty-second Street, PO Box 3861, New York, NY 10163-3861. This organization will prevent your name from being given to any of the large mailing list companies. You can write directly to companies from which you receive such mail, and they will remove your name from their mailing lists. Or you can recycle the junk mail. Whether you open it or not is still up to you.

Another of the fifty simple things you can do is "snip six-pack rings." In one day along the Texas Gulf coast, volunteers picked up more than fifteen thousand of the plastic container holders with which we are all familiar. These can become death traps for numerous sea birds and other marine wildlife, yet people continue to discard them on beaches and out of boats. The consequences include strangulation and drowning of gulls, terns, pelicans, seals, and sea lions. The book suggests cutting each of the rings before discarding the plastic holders. Also, pick up any you see and snip them before you put them in an appropriate disposal area. Even those that end up in landfills can entangle wildlife that frequent them. The most straightforward solution would be to support a prohibition against production of six-pack rings.

134

Junk mail and six-pack rings are just two examples of the barrage of environmental impacts over which we have some control. Others include the use of styrofoam cups, the excessive use of paper and other products from trees, and the overuse of energy associated with oil or electricity. We can take individual action by reducing our personal consumption. We can also support campaigns that discourage production of environmentally harmful products and encourage ecologically sound alternatives.

But environmentally positive undertakings often have a downside. Someone has to pay. Some businesses would not profit from certain recommended actions; these businesses tend to think poorly of suggestions to decrease, alter, or terminate production of their products. The counterposition maintains that environmentally destructive companies have been warned; they should take measures not to depend on products that degrade the environment we all depend on. A balance must be achieved between profit motives and what is best for the world's environments. Gradual phase-outs might be a palatable approach for solving problems in which abrupt termination of production could result in unexpected unemployment. Another approach calls for companies to develop ecologically acceptable products in place of those that despoil the environment. Consider also that when a shift in production is made, though some jobs may be lost, others may be created.

One of the most powerful environmental contributions an individual can make is to educate others about the importance of maintaining biodiversity and protecting natural ecosystems. One way to do this is to get your views into the open. Most communities have a newspaper that represents the voice of the people on the editorial page. Write a letter; maybe it will be published. Encourage others to write letters. A well-timed letter campaign might inspire community action. A friend of mine, Isabel Vandervelde, wrote a letter to the editor of a small-town newspaper with a theme that is being echoed around the country. One of her first sentences asks, "Isn't there any way developers can be restricted in their purchase and use of land?" Her final plea is, "Help! Someone stop this greedy march of senseless overdevelopment." Her message is applicable in many small towns and villages. We are losing natural habitats, wildlife, and small-town charm throughout the country by giving unrestricted license to commercial development.

The time has come for us to establish a tougher plan to preserve our environmental dignity. We need zoning laws that require all development projects to obtain approval of the community's citizens, not just those who stand to make a killing by selling their property. The profits to be gained (and their distribution) should be weighed against what will be lost. Everyone should be informed. In controversies over whether to keep natural habitats and peaceful neighborhoods or to develop yet another shopping area, let the community vote. My bet is that in most small towns and rural districts (and even in some cities) the majority of the residents would rather stroll through a park or woodland or down a tree-lined street than drive to a blacktop parking lot surrounding a mall.

135

An example of unnecessary development in my neighborhood is an apparently senseless transfer of store locations. An enormous grocery store was built in a former residential area. Tall pines and spreading oak trees were destroyed to create the new shopping area. But the consummate irony was that an enormous grocery store owned by the same company was already located across the street from the new store, about the distance of three, century-old loblolly pine trees laid end-to-end. That supermarket is now closed and empty. What's going on here? This doesn't even seem like good business.

I do not doubt that developers, tax accountants, lawyers, and politicians would point out that I do not understand commercial business development, franchises, leases, and other money matters. They would be right; I don't. But I do understand that the laws and procedures governing such transactions were created by people and can be changed by people. We can change the rules for commercial development. Perhaps we need a new business ethic that maintains that the preservation and protection of trees and wildlife are the first concern. Maybe commercial profit should be the second order of business, not the first. Whether it's a mall, an office "park," or a shopping center with a fancy facade, the final product is the same—concrete buildings. And the cost is the same—destruction of natural habitats, loss of native wildlife. The ultimate insult is commercial business's peace offering of scrawny trees in concrete pots.

The foundation of a democracy is that the majority rules. Popular vote should decide what restrictions to place on commercial development. Democracy also entails electing officials to represent us. Any sizable town, along with every county and state, has individuals put in office by majority vote. Inform those whom you have elected that you want a reassessment of zoning rights. Consider supporting laws that require identifying individuals who make a profit from a major commercial development. Let your representatives know how you feel. Ask them what they are doing to replace unbridled personal profit for a few with the preservation of the environment for the majority.

The "someones" who can stop "the greedy march of overdevelopment" are we ourselves. A letter fashioned after the following should get the attention of most elected officials.

Dear [Elected Official]:

I am writing to protest the impending destruction of a woodland in our neighborhood for the building of yet another shopping center that we do not need. Some of the trees that will be destroyed are older than anyone in our town, and a few are older than the town itself. As a taxpaying voter, I request that you inform me of the procedures to identify who will profit from this commercial venture. I already know who will lose. Most of us who live here. I would like to have the profit makers justify publicly to our community why the destruction of a peaceful woodland and its attendant wild animals and plants will serve us better than their preservation.

Please do not respond by telling me that the shopping center will create jobs for the unemployed. Many of those jobs will be taken by people from elsewhere who move into our town. Most of us who live here will be satisfied with the status quo of a quiet little town with trees.

Also, please do not tell me that the shopping center is being built on private property that has been legally purchased. Instead, advise me about what you are doing to change laws that allow the unharnessed annihilation of woodlands and wildlife. The fact that a forest happens to be on a particular piece of real estate should not automatically permit its destruction. In addition, the destruction of even a small forest can cause watershed problems that extend beyond the boundaries of the owner's property and affect all of us. I have no quarrel with private ownership of property; private lands in the right hands are generally the best cared for. But land targeted for commercial development will not serve the best interests of your constituency.

I realize that I must compete for your attention with lobbyists who support the shopping center and probably contribute substantially to your re-election campaign. The strength of my petition is that I have asked several others in your district to join me in this request. Their names, addresses, and signatures are given below. We may contribute less money to your campaign than corporate donors, but we will have a lot more votes on election day.

We look forward to your timely reply.

Sincerely,

You might ask your newspaper to run a copy of the letter. You may find that many others in your community favor environmental protection for everyone versus financial profit for a few.

If we can develop the basic understanding that native plants and animals and their habitats are part of our heritage, we will make progress. One way to do this is to establish symbols. Every schoolchild knows that the bald eagle is the national bird. Many probably know the rose is the national flower. But whoever heard of the national insect? We could have a national insect, if a proposal by the Entomological Society of America were accepted. To have such a proposal approved by a cautious Congress will require the support of individuals throughout the country.

The insect recommended for this honor is the monarch butterfly. This probably doesn't seem odd to natives of Idaho, Illinois, and Vermont, where the monarch is already recognized officially as the state butterfly or insect. But to many people, especially those who regard insects as pests, the idea of a revered insect may sound strange.

Most people know what a monarch looks like. The adults have burnt orange wings that sport black lines and a heavy black border with white spots. The two-inch long caterpillar has a series of black, yellow, and white rings. The caterpillar eventually goes into a resting stage and forms a cocoonlike chrysalis that is lime green with golden spots. About six weeks later the fully developed adult butterfly emerges.

For several ecopolitical reasons, I wholeheartedly support the idea of having a national insect. Insects get a lot of unfavorable attention from the actions of a few renegades such as fire ants, killer bees, and gypsy moths—none of which is native to North America. We need to identify a species like the monarch to represent North American insects, the vast majority of which have no negative impact on humans. Many, indeed, have positive impacts. Perhaps designating a national insect will inspire adults as well as schoolchildren to learn something about the life cycle of the monarch and will create a better understanding of the ecology of this and other insect species.

The monarch is an excellent choice for a national insect. It is beautiful to behold and is admired throughout the countries of North America. Like many other insects, the monarch pollinates the plants we appreciate. Unlike most other insects, it can fly through a flock of birds without being eaten.

The reason monarchs do not fall victim to hungry birds is that they taste awful. Monarchs lay their eggs on milkweed plants, which in turn become food for the caterpillar larvae. Milkweed produces a chemical that does no known harm to the caterpillar and that persists in the adult butterfly. The distasteful chemical keeps monarchs off the fast-food menu of almost all birds. Monarchs often drift through the air in a lackadaisical manner that suggests they know they are safe.

These beautiful butterflies perform a feat practically unheard of in the insect world. In autumn they migrate more than a thousand miles to overwinter along the coast of California or the mountains of Mexico. Some are known to have traveled more than eighteen hundred miles, including six hundred miles across the Gulf of Mexico! If you translate the four-inch wingspan of a monarch into airplane dimensions, Lindbergh's *Spirit of St. Louis* could have flown the distance from New York to the moon. Monarchs can also fly at high elevations, some being reported above three thousand feet.

Their ability to migrate, traveling farther and visiting more countries in a butterfly lifetime than most Americans do in their own, is another reason for showcasing this species. Plus, the monarch makes the journey without having done it before, guided unerringly by the marvel of what presumably is genetically programmed behavior. Monarchs clearly demonstrate the reality that the environmental conditions of one country or region can severely affect those of another. Destroying the monarch's ancestral wintering grounds in California and Mexico could mean that people in most of the United States and Canada would see no monarchs the following year. Establishing a species like the monarch as a symbol would aid in the development of a stronger national awareness of just how precarious the existence of many species really is. In any case, it would make a statement that all wildlife, including insects, represent an important, vital part of our natural heritage.

I predict some resistance to the proposal to have a national insect. Some people will see it as a means for environmentalists to strengthen attempts at preserving certain habitats. Imagine a community combating plans to ruin a habitat because the national insect would be destroyed. Such environmental attitudes, expressed nationwide, might make it difficult for opponents of wildlife preservation to resist conservation efforts openly. A national insect just might end up saving natural habitats. For the monarch butterfly to achieve that status will require the solid backing of many citizens.

Clearly, the worldwide loss of biodiversity, the ever-increasing threat of extinction of species, and the destruction of natural habitats are menaces for which no individual has a panacea. Each of us is like a leukocyte, a white blood cell, in a human body with an infection. Each leukocyte does its part by destroying one cell of the menace. Collectively, operating with a common goal, they destroy the source of the illness, returning the body to full health. Our problem is that many who might work as leukocytes to restore the environmental health of the earth are inactive. Perhaps they feel that a powerful antibiotic is the only effective remedy. But we do not have time to wait for an environmental antibiotic that may never be administered. Each of us needs to take some individual action immediately.

Choosing an
Environmental
Organization

The number of organizations devoted to environmental protection, preservation, and education makes a strong statement about the high value that the United States population places on the quality of natural environments. But the abundance of such associations can be daunting if you're trying to decide which one to support. Is the group taking a position you sanction and using approaches you approve of? With the diversity of environmental goals and missions today, the ranking priorities and developing strategies of the groups vary considerably.

In 1900, obtaining a comprehensive list of the world's environmental organizations and their objectives would have been relatively easy. At that time, North America still had more passenger pigeons living in the wild than it had societies working to preserve native wildlife. Half a century later, the number of environmental organizations had increased slightly and included the National Wildlife Federation, begun in 1936, and Ducks Unlimited, Inc., begun in 1937. Nevertheless, these and others were in the fledgling stages when California condors still soared above the Sierra Nevada. By the time the dusky seaside sparrow expired in 1987, chartered environmental societies and regional groups had reached almost uncountable numbers.

In addition to dozens of prominent and well-publicized environmental organizations such as Earth First! and Greenpeace, thousands more had been added to the list. Local nature groups and garden clubs had turned their attention to environmental issues. Organizations whose members were interested in a particular group of plants or animals were becoming established in many states and larger cities. The trend toward environmental preservation and conservation was growing. The timing was none too soon.

A heightened interest in environmental matters was developing on a national and international scale. The variety of action groups to choose from increased yearly, and the number of environmental journals burgeoned. *Buzzworm,* a slick magazine with striking photographs, appeared on the market in 1988, one of its objectives being "to explore and question the complex problems and solutions of worldwide environmental survival." The spring 1989 issue of *Buzzworm* gave a comprehensive accounting of many of the United States organizations with environmental or conservation goals. The sample included most of the major organizations and numerous minor ones that seemed to be doing well. Most of them still are. *Buzzworm* recognized 142 organizations that ranged alphabetically from the African Wildlife Foundation to Zero Population Growth. All of them published magazines or newsletters promoting conservation efforts.

The members of some environmental organizations concentrate their efforts to assure the survival and well-being of selected species or taxonomic groups. Birds and mammals seem to be the most popular animals to protect. Some organizations are dedicated to particular mammals, including bighorn sheep, bats, and bears. At least two have focused their efforts on preserving wild horses in the Southwest, though some would argue against such efforts because horses are not a native American species. The larger of these is called Wild Horse Organized Assistance, or WHOA. Bluebirds, loons, and trumpeter swans all have organizations committed to their welfare. Six groups concentrate their efforts on hawks and other raptors. Even reptiles have their supporters in the Desert Tortoise Council. Plants are not excluded. Two groups work toward the preservation of wildflowers, and the goal of the Save the Redwoods League is obvious.

Some environmental organizations, such as the World Wildlife Fund and the Sierra Club, have a global perspective toward habitat protection. The American Littoral Society is concerned with the conservation of coastal areas, and the American Cave Association's purpose is to protect caves. Others, such as one to preserve the Indiana Dunes and another to promote an understanding of the Chihuahuan Desert, have a limited geographical scope.

Although some environmental organizations have special missions, virtually all strive to protect and preserve habitats or biodiversity. The American Cetacean Society and the Pacific Whale Foundation are dedicated to saving the ocean environment as well as to protecting whales and dolphins. The Peregrine Fund, initially directed toward recovery efforts for peregrine falcons, supports protection for all rare and endangered birds of prey. I was pleased to find out that the International Primate Protection League maintains a sanctuary for gibbons.

Some organizations, although environment-spirited, have a secondary goal in their plan to preserve and promote a species. I think we all have a pretty good idea what the National Wild Turkey Federation, Quail Unlimited, and Ducks Unlimited, Inc., have in mind for these species. Nonetheless, such organizations can be extremely important in preserving natural habitats.

The membership levels and bank accounts of some of the larger environmental organizations are impressive. Greenpeace, with the goal of "protecting and preserving the environment and the life it supports" was reported by *Buzzworm* in 1989 as having one million members and an annual income of more than $26 million. Greenpeace now claims more than 1.8 million members and an annual budget of $44 million. The National Audubon Society, whose goal "to conserve plants and animals and their habitats" doesn't sound much different from that of Greenpeace, has more than 600,000 members. Another with a similar goal is the Nature Conservancy, with more than 500,000 members and an annual income of about $1.7 million. The Nature Conservancy seems to set the record among environmental organizations for bringing in donations that result in land acquisitions to set up nature sanctuaries.

Another environmental journal, *Outside,* published an article in September 1990 entitled "Inside the Environmental Groups." A banner on the cover asked "Who's *Really* Saving the Earth?" The article is a user's guide to the publication's "Top 25" of more than five hundred environmental groups they recognized. The ranking scheme is no less rational than those for college football or basketball standings, and the facts make for some interesting comparisons.

A description of each organization includes a synopsis of its background and goals. Categories for each are "Who's in charge," "What they've done lately," "Where the money goes," and "What they don't tell you." Each group is also given a rating from 1 (Milquetoast) to 5 (Bombthrower) as an index of activism. For example, the

Milquetoast rating is given to Ducks Unlimited, Inc., and the Nature Conservancy; the Bombthrower rating is given to Earth First! and Citizens Clearinghouse for Hazardous Wastes. The National Audubon Society gets a 2, Sierra Club a 3, and Greenpeace a 4. Greenpeace has the biggest membership. The National Wildlife Federation is a distant second with almost a million, only a few thousand ahead of the third place contender, the World Wildlife Fund.

One important consideration when donating to any charity or cause is how much of the budget is used for overhead costs. That is, how much of each dollar is used to finance the organization's programs directly, and how much goes to support personnel in the home office? Among environmental groups, no relationship is apparent to me between budget size and the proportion that goes into overhead. For example, 34 percent of the $4.6 million Defenders of Wildlife budget goes for overhead; Conservation International has a 15 percent overhead with the same budget. The lowest overhead rates in the "Top 25" are about 14 percent (National Wildlife Federation and National Toxics Campaign). The highest ones (30 percent or more) include the National Audubon Society and Sierra Club. *Outside*'s figures reflect numbers provided by each organization.

The article gives an often entertaining view of a "typical member" of each society. For example, a Greenpeacer "has lasting regrets about not skipping class to attend antiwar rallies," whereas a National Audubon Society member "feels a twinge of guilt whenever she dresses the Christmas turkey." The Environmental Defense Fund member is portrayed as a "lawyer with a green conscience and a red Miata." The members of Defenders of Wildlife are typified by "a bearded, middle-aged man who likes following game trails through the underbrush, wearing knobby Vibrams and no underwear." The descriptions portray group personality, an important consideration when choosing a group to join or provide support for.

As you might expect, environmental groups often compete for certain resources (such as members) and are not one big happy family that agrees on everything. The *Outside* article gives some insight into some of the not-so-friendly relationships. For example, Friends of the Earth was founded in 1969 following a political split within the Sierra Club. Conservation International is a splinter group from the Nature Conservancy. And the lawyer-run Sierra Club Legal Defense Fund, an offshoot of its namesake, takes pains to point out that it is not the Sierra Club.

The larger, generalist environmental groups have broad objectives. But some groups aim at very specific targets. The African Wildlife Foundation focused on the elephant ivory problem long before it became a fashionable concern. American Rivers, with fourteen thousand members, has a fondness for water moving downhill and an aversion to man-made dams. An objective of a group called Environmental Action was to identify a "dirty dozen" congressmen whom they accused of selling out the environment.

144

One low-profile group not mentioned in either magazine article is the National Fish and Wildlife Foundation. This not-for-profit organization, established by Congress in 1984 and based in Washington, D.C., matches private contributions with federal funds. The foundation's goals include habitat protection, policy development, research, and resource management. It reports one of the lowest administrative budgets of any group of this nature, carrying out operations at an overhead rate of only 3 percent. If you want to get the most for a contribution, this may be the route, although the foundation prefers to receive sums of one hundred dollars or more, to reduce handling costs.

The tactics of some environmental groups make them controversial. Anyone who reads newspapers or magazines or who watches TV has heard of Greenpeace. But ask what people think Greenpeace is or does. The answers are varied and often vague. Some people have an image of Greenpeace from a recollection of one particular incident or another (Didn't the French secret service bomb a Greenpeace ship in New Zealand? Didn't the group expose to the world how baby harp seals are killed for the fur trade?). Our perception of an organization is often based on a single incident. Opinion on whether the organization is beneficial or detrimental to society can be influenced by the depth and manner of news coverage. Newspapers and television often give a fragmented overview and sensational accounting, and many magazines tend to editorialize. I contacted Greenpeace directly to learn exactly what its goals are and how it goes about achieving them.

The organization was established in 1971 in Canada in opposition to nuclear testing on Amchitka Island in Alaska. Today Greenpeace focuses on three major areas: pollution by toxic chemicals and radioactive materials in the environment, destruction of the world's oceans, with emphasis on marine mammals, and depletion of the ozone layer and the threat of global warming. Can anyone oppose the stated goals of Greenpeace? Do you know anyone who thinks that environmental contamination is something we should be trying to accomplish instead of trying to stop? Obviously not. Have you ever seen a sign that says "Do your part to destroy the ozone layer"? Absurd. Do you hope to see a dolphin the next time you are on the coast, so that you can kill it? Of course not. Why, then, do we not all belong to Greenpeace?

To stop what it perceives as injustices against the environment and those living in it, Greenpeace is dedicated to confrontation and to taking action without committing violence but often at some personal risk. The paths to achieve these goals are many. As they themselves report, Greenpeace supporters have used inflatable boats to interfere with large commercial ships engaged in killing whales or seals. They have blocked smokestacks and liquid waste pipes that were discharging chemicals into the air or water. They have publicized in a variety of ways what they judged to be environmental abuses, based on what they felt was sufficient documentation. Several specific objectives on the Greenpeace agenda include curtailing whaling by Japan and Iceland, and making Antarctica a world park. The organization also wants to ban the use of large drift nets in the ocean and to work through public and political pressure to stop nuclear testing by the military. Greenpeace itself states that its supporters "are determined to continue until the world is a safe and healthy place for all its inhabitants."

Some people do not approve of Greenpeace's tactics; some disagree with specific aims. Each of us, based on our background, training, and place in the world, has a personal approach to dealing with issues that confront us. To achieve a long-term goal, innumerable short-term goals must be met. Some of us may choose a short, direct path, but one with potential pitfalls. Others prefer a circuitous route with fewer risks, except, of course, the risk of taking too long. The urgency of the situation may affect which path is chosen. To reach a dying relative (or dying planet) may require a direct and potentially hazardous route. Thus, while we may all endorse Greenpeace's general goals, membership and financial support will be right for some but not for others. Unless you already know all of Greenpeace's goals and methods, why not get some more information—from the organization. The same approach should be used with any group you feel might be appropriate for you. Write it for facts about its goals and positions on environmental problems that concern you. Find out how it achieves its goals.

For example, the Rainforest Action Network (RAN), dedicated to the preservation of tropical rain forests, distinguishes itself from other environmental groups set on tropical rain forest protection by an emphasis on grassroots education and an activist approach to dealing with this serious problem. The destruction of tropical rain forests is one of the hottest environmental topics of our times. We have all heard that rain forests are being destroyed at a fast and disturbing rate. We have an idea of who some of the culprits are. We have been told that we should do something to stop the devastation.

146

Like many other environmental groups these days, nonprofit or not, RAN can sell you colorful T-shirts and pertinent books. But it also delivers a powerful message about the fate of the rain forests of the world and provides guidance for what the average citizen can do. It also answers many commonly asked questions about rain forest destruction. And if you are already convinced that tropical rain forests are worth protecting from the ravages of commercial exploitation, RAN offers suggestions for how you can help, including continuing to educate yourself about the rain forest issue. You be the judge on whether you want to heed the other suggestions.

Among those suggestions is RAN's recommendation that you not buy furniture or paneling made from tropical trees such as rosewood or mahogany. This will not be an imposition on most of us. It also advises against using disposable chopsticks, because the wood is usually from tropical trees. How many trees does it take to provide the world with chopsticks? Consider the number of Oriental restaurants in the world. Each day they throw away countless chopsticks—millions of pieces of tropical wood.

Another RAN recommendation is for people to stop eating fast-food hamburgers or processed beef products until a law is enacted requiring that all beef products be labeled with the country of origin. Beef can be raised more cheaply in tropical areas, but doing so requires clearing rain forests. Advocates of beef labeling maintain that many companies using tropical beef do not admit it. A labeling law would spotlight the use of beef from tropical countries. My bet is that most Americans are not ready to give up hamburgers completely, but many would be willing to pay a little more for them and eat a few less. If you want to express concern about the use of tropical beef, RAN suggests letting the United States secretary of agriculture know you are in favor of a beef labeling law that follows the beef all the way to the consumer.

RAN also suggests you write to the World Bank. According to RAN, the World Bank uses our taxes to finance the building of dams in rain forest areas, putting the tropical countries in debt to Japanese and United States banks. Such dams destroy millions of acres of forest and displace jungle tribes from their homes. Like most dams, even in this country, the projects are considered environmentally destructive, pork barrel boondoggles.

RAN states that cancer-causing herbicides and pesticides banned in the United States are used extensively in rain forest countries. Although specific plants or insect pests are the targets of these toxic chemicals, their indiscriminate use damages other organisms. Further, the cancer-causing properties are still there when tropical foods are shipped to other countries, including the United States. Apparently, some United States companies reap a profit from the sale of these chemicals, with the sanction of the U.S. State Department. RAN suggests you write your senators and representatives to protest this activity.

Another environmental organization, Earthwatch, is entirely different from the others, with a positive approach that almost everyone approves of. Earthwatch provides the ultimate form of environmental education and is a one-of-a-kind organization that anyone with environmental interests and a yearning to do ecological research should come to know. A nonprofit institution, Earthwatch has a well-defined and admirable goal: funding research. Promoting environmental education through volunteers who are personally involved in exciting field research projects is a side benefit.

Volunteers pay for their opportunity to be involved, but the costs are less than the expenses of a typical two-week vacation, and the experience is much more meaningful. Each year over three thousand volunteers known as the EarthCorps assist research teams in collecting scientific data. Volunteers benefit from direct participation in a research project and dealing firsthand with nature. Researchers benefit from enthusiastic field assistants. Earthwatch gets its reward from making things happen.

A typical issue of *Earthwatch Magazine* gives a sampling of research projects that can accommodate volunteers. Some are of cultural or historical interest, but many Earthwatch projects involve ecological research. Opportunities abound worldwide. For example, a project in Australia addresses the question of how effective the fast-disappearing native vegetation is as a refuge for the wildlife dependent on it. An Earthwatch volunteer can expect to walk up to ten miles a day capturing and censusing kangaroos, spiny anteaters (an egg-laying mammal), and native birds.

Volunteers who conduct such surveys contribute to our general knowledge of the environmental state of affairs in western Australia.

A Central American project is conducted on a coral reef, the second largest in the world, off the coast of Belize. One question being asked by the researchers, amid one of the most pristine and beautiful reefs left in the Caribbean, is how organisms partition the resources of a coral reef habitat. Among the various species of reef fishes are ten kinds of moray eels, which as adults have no known natural enemies.

I have been involved in an Earthwatch project in the salt marshes of Kiawah Island near Charleston, South Carolina. The research, conducted with my coinvestigators Tony Tucker and Jeff Lovich of the Savannah River Ecology Laboratory, is a study of diamondback terrapins. The project is environmentally significant because diamondback terrapins, the only North American turtle that lives exclusively in coastal salt marshes, were seriously overexploited as a luxury food item early in the century. Terrapin stew was a delicacy, and millions of these turtles, especially the large females, were removed from marshes and estuaries for restaurants. Whether terrapin populations have fully recovered from the assaults is uncertain. The Kiawah project investigates the ecology of an unusual species. Earthwatch volunteers learn the research techniques used to discover ecological facts about turtles and experience the wealth of ecological phenomena of barrier islands and the Atlantic Coast salt marsh system.

All Earthwatch projects are adventures, and the exercise can be strenuous. Some offer unpredictable excitement, as with three projects conducting surveys of humpback whales in Hawaii, New England, and Baja, Mexico. Observations are made either from motorboats or from shore stations. Another project studies orcas (killer whales), the largest dolphins in the world, aboard a catamaran in Puget Sound. Programs such as those offered by Earthwatch are of untold value in developing an appreciation of natural habitats, establishing an awareness of the difficulties and excitement of field research, and promoting environmental education through personal involvement.

The diversity of today's environmental groups is remarkable. True, each has certain failings and limitations; each has ways it could improve. But collectively they make a powerful statement: Americans are concerned about the environment and are willing to donate their time, energy, and money to make the world a better place in which to live.

Many people are deeply concerned about the loss of biodiversity and habitats, about uncontrolled pollution and waste, and about commercial enterprises and politicians who equate progress with environmental loss. The world's environmental organizations are making powerful and positive inroads into our environmental thinking and culture. The options for a concerned citizen are many. If you want to work with others, identify a group that suits your environmental goals. Or, if you can't find just the right group, start a new one based on your own ideas.

Biotechnology: A Cure
or a Band-Aid?

I recently encountered an approach that some see as an alternative to the removal of animals from the wild. Without moving anything but my eyes, I could see a thirty-five-pound iguana lizard, a ten-foot python, and more than a hundred red, yellow, and black king snakes. The situation was one that any comfort-loving herpetologist would enjoy. I was not standing in a tropical jungle; I was in the lobby of a Florida hotel. You could tell it was a nice hotel—the iguana was on a leash and the king snakes were well-cared-for babies. The python was a healthy albino unlikely to live long in any jungle.

Tony Mills, of the Savannah River Ecology Laboratory, and I were attending a meeting in the hotel, across the street from Universal Studios in Orlando. We never went to the studios that weekend; the best buy for excitement was in the hotel itself. It was the site of the Reptile Breeders Expo, sponsored by the Central Florida Herpetological Society. You could see a thousand pythons and boas, two thousand king snakes, and three thousand registered human participants. You could also see a wide variety of genetic stock in all categories. The snakes represented what is becoming a fast-growing hobby in the United States—a hobby seeking recognition as a reputable part of the national pet trade.

Expo guidelines required that all animals on display be healthy, all cages be clean, and all wildlife laws and regulations be strictly adhered to. Attendants included three veterinarians provided by the expo to examine specimens. Wildlife officials were present to check permits. Only captive-bred specimens, those born and raised in captivity, were allowed. In other words, reptiles captured in the wild were not welcome, nor were their captors. This may seem like a difficult rule to enforce. How can you tell if a snake has been caught in the wild or is captive bred?

During the setup on the first day, Tony and I noticed two adult coachwhips, a species of snake native to Florida and other southern states. Anyone who deals extensively with snakes knows this species is highly unlikely to be bred and raised in captivity. Because they do not ordinarily make good pets, they are not a popular pet trade species. The chances that someone had hatched coachwhip eggs and had the patience and know-how to raise the young to adulthood were slim. The next day the coachwhip exhibitor was escorted from the display hall; the presence of a wild-caught snake had been detected by expo officials and was not tolerated. I saw another exhibit removed because the owner did not have the proper state permit needed to sell or exhibit reptiles.

Although captive breeding is controversial at many levels, some believe that breeding reptiles in captivity for the pet trade is a form of biotechnology that could mitigate the loss of some species in the wild. The rationale is that making a species available to the pet trade through captive-bred stock eases the pressure to capture the species in the wild. Captive-raised animals are often more desirable as pets because they are healthier and free of parasites. Also, some species of snakes can be programmed to be better pets if their feeding patterns and preferred foods are established at an early age.

However, opponents question whether captive breeding for the pet trade is truly a form of conservation of wild animals. In addition, some claim that development of a pet trade actually creates a market, thus encouraging the capture of wild specimens. This criticism directed at selling captive-bred reptiles is one that haunts all regulated trade involving wild animals. The argument is that when only animals born and raised in captivity can be sold legally, the result is poaching, smuggling, and illegal importation of specimens caught in the wild. The counterargument is that such illegal activities are most likely to occur when it is cheaper to catch an animal in the wild than to raise one to the same age or size. Once it becomes appreciably cheaper to raise a species in captivity, illegal practices involving wild capture dwindle. The issue is difficult to resolve. Regardless of the enterprise, some people simply do not play by the rules.

The reptile trade has its special problems because of the nature of the animals themselves. One criticism is that reptiles acquired as pets are often ill-housed, poorly fed, or otherwise mistreated, mainly because most owners do not know how to care for them. Ignorance associated with keeping reptiles as pets need not be a lasting problem. A properly handled exhibition such as the Reptile Breeders Expo can provide ample opportunity for pet owners to learn how to care for their investments.

Among the exhibitors at the Orlando expo were those selling mice for snake food (you can order them frozen in any size), snake cages with special heating compartments for the inhabitants, and hundreds of books and pamphlets on proper housing, care, and feeding. What a reptile pet owner had not learned from experience could be easily found in one of those sources. Not all reptile owners may yet know about proper care, but they certainly have the means to learn. And the argument that people will not care properly for snakes and other reptiles, even if they know how to do so, is not valid. The pet owner, not the type of pet, determines how well cared for an animal is. Unsavory examples of improper care can still be found for the most popular pets around, dogs and cats.

Reptiles are not the only exotic animals acquired as pets. Hundreds of thousands of parrots, cockatoos, parakeets, and other exotic birds are legally and illegally imported into the country each year. From 1984 to 1988 the United States is known to have imported 3.5 million exotic birds for the pet trade. It seems ironic that we have more environmental groups than any other country yet serve as the world's largest pet market for tropical birds.

Each year, as tropical rain forests disappear, the world becomes less green. But even where jungles remain intact, the reds, yellows, and blues are gradually being eliminated, as exotic birds such as rosellas, lorikeets, and macaws fall victim to the pet trade. Many end up as healthy, maybe even happy, pets. Many others die between their homeland and someone else's. Whatever their final fate, they are no longer part of the jungle. Removal of these animals from their native habitats is an unhealthy environmental practice. Many are threatened species. Many of those not yet identified as such cannot sustain continued removal. Nonetheless, some countries still permit their capture and export.

We can determine how many birds are imported legally; no one knows how many are smuggled in illegally. But the number is sure to be high. Because bird smuggling must be secretive, many illegally transported birds die from malnutrition, exposure to extreme temperatures, and even suffocation. One estimate is that more than thirty thousand birds die each year as a result of the illegal pet trade. That's more than one bird every twenty minutes.

Solutions for the loss of bird life from the wild and the welfare of birds in the pet trade are of major concern to the World Wildlife Fund (WWF). Many people recognize WWF's little panda logo. It has given high profile and priority to passage of congressional bills to control the import of wild-caught birds from other lands. One WWF-supported bill calls for phasing out all legal imports of wild-caught birds over a five-year period and encourages captive-breeding efforts. The essence of WWF's solution is that if successful captive-breeding programs for exotic birds could be established, pet lovers would be able to have their birds, and the jungles would be able to keep theirs.

Another bird trade bill put before Congress at the same time calls for an immediate ban on imports of any wild-caught birds. The intent of the bill is to bring the exotic bird trade to a screeching halt. WWF does not support the bill—a good example of the conflict among supporters of environmental reform. Both bills have the same objective, but the paths to reach it are different. As the president of WWF put it, the WWF-sponsored Exotic Bird Conservation Act is designed to build consensus among all concerned parties by addressing both the environmental and economic aspects of wild-bird trade. The demand for exotic birds as pets is great and cannot be sustained by captive-breeding programs now in place. An immediate ban on imports of wild-caught birds would not provide sufficient time to establish such programs.

The WWF claims that the other bird bill would drive the exotic-bird pet trade underground so that even more would die in illegal shipments before they reached their cages in America. People would still seek birds as pets, but they would get most of them illegally. Perhaps the conflict over the two exotic-bird conservation bills is a positive sign: An environmental issue has congressional representatives quarreling over which legislation will more expeditiously solve the problem. Let us hope that at least an effective compromise bill can be passed before it is too late for the recovery of all species.

The pet trade is not the only controversial activity associated with captive breeding. Differences in opinion also exist about the approach of combining captive breeding with the reintroduction of animals into the wild. Patricia J. West, in

Encyclopaedia Britannica's *1992 Yearbook of Science and the Future*, describes what has become this last resort for the preservation of some endangered wildlife species. Individuals of a species, often the last survivors, are removed from the wild and bred in captivity. Once their numbers increase, if they do, they are reintroduced into their native habitat. The technique has been successful for some species. Three decades ago, peregrine falcons were truly a dying breed; only a flicker of hope existed for their recovery. They are still classified as endangered, but their numbers have

increased remarkably, in part because of this relatively new approach to dealing with species on the verge of extinction.

Today's unparalleled loss rate of species has created a sense of urgency, a feeling that desperate measures are needed to stave off disaster. Only recently has the technique of captive breeding and reintroduction been attempted on a broad scale with a diversity of species. Over the last two decades, zoos have become principal sites where species in danger of extinction can be safely kept, possibly bred in captivity, and then later released back into their natural habitats. The program with peregrine falcons is considered to have been among the most successful.

In the 1950s and 1960s, the eggshells of many birds became fragile as a result of pesticides that had accumulated in their prey. The survival of young peregrine falcons declined, greatly reducing their numbers worldwide. In 1942 more than 300 pairs bred in the eastern United States; by 1960 none were left in the region. In 1970 a captive-breeding program was started for peregrines, using animals from the western United States and Europe. The first birds were released in the eastern United States in 1974. As a result of the program, more than 175 wild pairs are expected to be breeding in the region by the early 1990s. Despite some successes, captive-breeding programs for purposes of reintroduction into the wild must not be viewed as the solution to our wildlife problems. No solution is possible without correcting the causes of the decline. Captive breeding and reintroduction is a treatment, not a cure.

Many reintroduction programs have been dismal failures. In an attempt to save the endangered Hawaiian goose, more than fifteen hundred captive-bred birds were released. Wildlife officials have yet to reestablish wild populations. Hunters and introduced predators are held responsible for the goose's lack of success. Another failure is the reintroduction of the Guam rail, a flightless bird. Successfully bred in captivity, the bird has not been reestablished on the island of Guam because of the presence of the accidentally introduced brown tree snake, which eats the bird's eggs and young.

Success in a reintroduction program depends on two vital elements: achievable captive breeding and sustainable native habitat. Such programs have resulted in the return of golden lion tamarins to Brazilian rain forests, endangered sea turtles to the Gulf of Mexico, and the Gila topminnow to its native habitats in the arid Southwest. But the costs of this approach to preservation can be high and the results uncertain. By the time each golden lion tamarin has been returned to its native habitat, it has cost more than twenty thousand dollars. As Nat Frazer notes, none of the fourteen thousand young Kemp's ridley sea turtles hatched in captivity and returned to the Gulf are known to have nested. And the Gila topminnow remains in a life-and-death battle with an introduced species, the mosquito fish, which has invaded the topminnow's natural habitats. We should not embrace the reintroduction concept without careful consideration.

The reasons for the high expense of some captive-breeding programs are obvious. Travel and field expenses are necessary for the initial capture of animals and for their subsequent reintroduction into the wild. Research is another expense. Determining how a species can best be bred and raised in captivity may take years. For some species, time and effort must be invested in training captives for life in the wild. In some cases, success has yet to be obtained, as with the whooping cranes in Idaho. In addition, captive species must be housed and fed, and some species need veterinary care.

According to Howard Hunt of Zoo Atlanta, a pair of endangered Morelet's crocodiles from tropical America requires a twenty-by-sixty-foot enclosure with a pond. From 1975 to 1978, almost one hundred young were hatched at the zoo and released in their natural habitat in Mexico. In this case, the routine costs of the breeding program were offset by zoo-goers' appreciation of the captive crocodiles.

Rearing some species in captivity requires unusual strategies. The black-footed ferret of Wyoming, presumably extinct in the wild but now maintained in captivity, was reduced to fewer than two dozen individuals. Among the approaches used to raise the rare mammal in captivity was the use of a Siberian polecat, a close relative, to serve as surrogate mother for young ferrets whose mothers were unable to nurse them. Such innovative approaches are common in captive-breeding programs, but their success depends on trial-and-error research that can be expensive and time consuming.

Critics of these federally funded programs are many. Taxpayers who are indifferent to the plight of endangered species condemn such efforts. Even some conservationists maintain that the money could be better spent in other conservation efforts. For instance, preserving habitats is likely to assist far more species and individual animals than are programs aimed at saving a single struggling species. An endangered habitat act has even been proposed, but never passed, in Congress. Its passage would ensure the preservation of both terrestrial and aquatic habitats, thus safeguarding the native species that live there.

One concern expressed by ecologists is that captive-breeding and reintroduction programs do not properly address the real problem. Such animal welfare programs deal with the short-term problem—too few animals. They do not deal with the problem that placed the species in jeopardy in the first place. Problems caused by pollution, habitat destruction, or hunting pressures are not solved by releasing more animals. Reintroduction of the peregrine falcon was successful only because a major cause of the bird's decline, pesticides, had been removed as a threat. Removing the environmental problems confronted by reintroduced species is vital to the ultimate success of these programs.

Another problem with captive-breeding programs is that they may foster the perception that the decline of animals in the wild is easily fixed. To imagine that reintroduced species will bounce back and be fine is dangerous; it lessens concern about the continued degradation of habitats on which the animals depend. Such programs may ease our environmental conscience, but they are of little long-term value unless we deal with the reason the problem exists to begin with.

Besides using captive-bred animals as the source for reintroductions, some conservation strategies involve the relocation of individuals of a species from an area of abundance to areas where they have been extirpated. Although relocation may be a solution for some problems involving species loss, C. Kenneth Dodd, Jr., of the U.S. Fish and Wildlife Service and Richard A. Seigel of Southeastern Louisiana University urge that a careful review of successes and failures be undertaken before it is advocated as an acceptable management and mitigation practice. Dodd and Seigel list two dozen cases in which species of reptiles or amphibians were the subjects in relocation programs, including ones in India, England, South Africa, North America, and the Seychelles. Of these, only five relocations are known to have been successful; six have clearly failed. The success of fifteen is still undetermined and may not be known for several years.

Dodd and Seigel offer several recommendations for the development of sound relocation programs; many of these recommendations will require extensive research. For example, the biology, including the genetics and social structure, of the species and the ecological constraints within the habitat are not always easy to determine. They recommend detailed studies to understand the causes of the decline of a species in a habitat. And long-term monitoring programs are essential, so that the success or failure of relocation can be determined. Relocation of animals may be an effective method for replacing species native to a habitat, but a sincere commitment to research will be necessary before we really know what we are doing.

A successful captive-breeding program cannot be started simply because we realize a species is in trouble. Captive breeding, like other biotechnological advances, requires trial-and-error research. The combination that unlocks the door to successful breeding in captivity is essential, and it is not always easy to determine. Each species is unique in its reproductive behavior, following guidelines established throughout its own evolutionary history. Time is needed to ensure success with most species, and with many, effective breeding in captivity has never been achieved.

In addition, most programs focus on species that appeal to us because they are cute, large, or furry, an interesting basis for deciding which species live or die. In the medical profession, the triage approach is used in a catastrophic situation to determine who should receive medical assistance. High priority is given to those who need immediate help and have a good chance to recover if they receive aid. Low priority is given to those who either will recover without help or will probably die anyway. The funds for conservation efforts are limited. Should we now place our efforts primarily on species where recovery is assured? Should we make a conscious decision not to expend resources on a life-support system to keep a doomed species alive? Some view as a poor investment money spent on species such as the California condor that apparently can no longer subsist on its own in the wild and is at the mercy of captive breeding.

The expense (and accompanying controversy) involved in captive-breeding and reintroduction programs may have a certain value in creating an awareness of the problem. The high costs of captive propagation should make us rethink the price of forcing species to flirt with extinction. Perhaps the costs will also force us to implement more-effective means of preserving wildlife. We need to focus our energies—and our conservation budgets—on protecting natural habitats and the animals that successfully live in them.

Removal of native species from a habitat can cause damage difficult to repair. The addition of the wrong species to a region can also cause severe environmental harm, often to an entire habitat; indeed, because of introductions of non-native species, some habitats may be seriously threatened. Solutions for how to manage animals or plants that have become established in a habitat where they do not belong are seldom easy. But resourceful scientists seem to find no end to clever biotechnological approaches for solving some of our species problems.

Feral animals—animals once domesticated but now wild—are a major concern in some regions. And as William D. McCort of Pennsylvania State University learned from a population of feral donkeys on a coastal island, controlling feral animals sometimes requires ingenious approaches. Compared to some feral introductions, such as wild horses or burros in the western United States, the problem was small-scale. However, the study demonstrates that basic ecological understanding of any species is critical and often not as simple as first surmised.

Ossabaw Island, located near Savannah, Georgia, is the third largest of the barrier islands along the Georgia coast. Through the efforts of Eleanor Torrey West, former owner of the island, Ossabaw has not been commercialized; it is the state of Georgia's first Heritage Preserve, dedicated to "natural, scientific and cultural study, research and education, and environmentally sound preservation, conservation and management." Setting aside or donating land for wildlife preservation is one of the greatest contributions a private landowner can make.

Wildlife abounds on Ossabaw: white-tailed deer, turkeys, alligators, cotton-mouth moccasins, diamondback rattlesnakes, pileated woodpeckers. Herons and egrets nest each spring in the island's two rookeries. During summer, loggerhead sea turtles return to Ossabaw's unlit beach to lay their eggs. The island is truly a sanctuary for native wildlife. But in addition to the wild animals, populations of three feral species lived on Ossabaw when McCort's study began: feral cattle and swine, originally left by Spanish explorers, and feral donkeys. Feral animals can become a problem if left unmanaged. They often compete with native wildlife for food and cause environmental destruction if they become too numerous.

Eleven descendants of Mediterranean donkeys, a gift from the island's owner to her son, were brought to Ossabaw and released in 1965. By 1975 the original population of eleven had increased to sixty-nine. Population increases were also observed in the feral cattle and swine on the island. If the island was to be returned to its natural condition, long-term population control procedures were needed. The cattle could be captured and removed to fenced areas on the mainland, and the pigs were trapped or hunted. But the donkeys were uncontrolled. Unlike the cattle, the donkeys could not be taken to the mainland; they were quarantined on the island because they carried equine infectious anemia, sometimes called Coggin's disease. The disease is transmitted by flies and mosquitoes and can be fatal to horses.

However, destruction of the donkeys was considered unacceptable, and other solutions were sought. Castration would have stopped reproduction but would have eliminated much of the donkeys' social behavior, which McCort was in the process of studying. After much debate, the decision was made to vasectomize all male donkeys—a zero population growth objective for Ossabaw Island. The idea was to allow the donkeys to live out their lives on Ossabaw with their social behaviors intact. The population would eventually be eliminated through natural deaths.

No previous attempts had been made to vasectomize a donkey, let alone an entire population. The Penn State investigators set out to capture all the donkeys. The ecologists soon learned that the animals were smarter and more agile than cattle; they repeatedly outmaneuvered attempts to capture them. Eventually, the researchers managed to trick them into feeding in fenced fields planted with millet, an irresistible treat for donkeys. After three months, all the donkeys were captured; two veterinarians performed the vasectomies.

Once the population control procedure was completed, the donkeys were allowed to return to their former freedom on the island. The researchers knew at the time that a few of the females were pregnant. Any male foals would be caught and vasectomized later to avoid more reproduction. A donkey foal takes two years to reach sexual maturity. Furthermore, according to McCort's earlier studies on the social behavior of these donkeys, young males must wait several years before they attain access to females for mating.

A female donkey normally gives birth twice over a two-year period. By the time she gives birth to a new foal, she has begun to ignore her older offspring. Younger males may stay in their family groups until they reach sexual maturity, but the older adult males, who consider the group females to be their mates, prevent them from mating. The younger males are unable to defend themselves against the more aggressive and experienced males. Eventually young males are forced out of the family group and join bachelor groups elsewhere on the island. By understanding the behavioral biology of the donkeys, the investigators knew that for the immediate future there would be little need to worry about a population increase. At least that's what everyone but the donkeys thought. Many factors can affect the way individuals of a particular species behave. Even though we understand the basic biological principles that govern behavior, we may still find exceptions, or discover we did not know enough. Scientists may create theories; the animals do not always heed them.

Two years after the vasectomies were performed, the donkeys delivered a surprise. The vasectomies had clearly worked; the adult female donkeys had not given birth to new foals. However, they had continued to protect their last-born young, which were no longer so young. So, the now sexually mature male offspring, with mothers providing protection, had not been driven out of their family groups. When the vasectomized adult males attempted to stop the young males from mating, their mothers fought off the aggressive attacks on their sons. The result was that young males were able to mate, and female donkeys were once again pregnant. Having learned a new ecological and behavioral lesson, the researchers were once again in the business of population control on Ossabaw.

Biotechnology can aid us in resolving environmental problems associated with habitat modifications and the decline in numbers of various species. However, many of our efforts are hampered because of limitations in our fundamental level of understanding about species. Biotechnological research, coupled with an understanding of natural ecosystems, could mean salvation for myriad species now experiencing crises worldwide.

We must give ecologists the opportunity and encouragement to conduct basic research. Research ecology is essential not only to advance biotechnological practices but to strengthen our basic ecological understanding of natural systems. Learning and understanding as much as possible about the world's organisms and environments will give us the ability to deal with both the expected and the unexpected. We will never know all there is to know, but we must continue learning as much as we can, while making use of the knowledge we already have.

Captive-breeding, reintroduction, and relocation programs may play an important role in helping us sustain the world's wildlife. But the best way to ensure that our descendants can enjoy the world and its wildlife is to preserve it before it is endangered. Everything else is just an attempt to rectify the problems we create.

Ecology Starts at Home

The message that even everyday animals can be the cause of intrigue and excitement was delivered to me by a flying squirrel, right in my own home. The southern flying squirrel is a fascinating little creature whose nighttime squeaks usually go unnoticed by the average person. But flying squirrels in the attic are hard not to notice; they are a nuisance. Add to the equation four children getting ready for school and complaining about how "the squirrels scrambled between the walls all night" and how they "heard them squeak all night" (always "all night"), and you can understand why I decided to set a trap.

Flying squirrels are cute. Being nocturnal, they have large eyes, alert and attentive. Attached to their front and back legs on each side of the body is a flap of fur-covered skin. When the legs are held out, the flap is extended and stretched flat, permitting the squirrels to glide from a high perch to a lower one. They cannot fly upward, but they can definitely glide through the air with the greatest of ease.

My wife, Carol, and the children would never sanction a trap that would harm or kill a flying squirrel. So I settled on a harmless metal box with slam-shut doors that biologists use to capture small mammals, and I used the universal bait, peanut butter. The trap worked. I caught a flying squirrel. Carol was getting the children ready for school when I emerged from the attic to show off the catch. Everyone— six humans and three four-legged pets—gathered at the front door, all intent on the scurrying inside the metal trap.

Kahlua was a fat black cat full of stealth and with a questful eye. For an instant, the stilettos at the ends of her toes were unsheathed. I knew she would like to demonstrate her predatory skills. Martini, on the other hand, was a vacuous-eyed cat, much loved for her inoffensiveness. Her predatory achievements consisted of two trophies. The first was a car-struck robin that had lain dead on the street for two days. Martini presented it to us on the front porch. That same year she proudly brought us a bright yellow maple leaf, dropping it gently at our feet. B. D. was a big, cuddly shepherd, the delight of babies and little children, but a fearsome watchdog. His repertoire of barks ranged from a meek-as-a-field-mouse whimper to a roar that could jeopardize mail delivery to houses half a block away.

I lifted the door of the trap to see the squirrel. Before I knew what was happening, we got our first ecology lesson of the morning: Flying squirrels can squeeze through tiny openings. From that point, things happened fast. The squirrel jumped from the trap and glided to a chair in the living room. Children squealed. B. D. barked. Martini bolted into a closet. And Kahlua made a graceful capture on the chair. Within seconds, Jennifer grabbed Kahlua, who released the squirrel unharmed. B. D. knocked down Michael and Susan Lane in an effort to get to the center of the action, and the flying squirrel scampered under the sofa. Everyone was shouting advice, and no one was listening.

All of us except Martini headed toward the sofa, and the squirrel made a hasty retreat into the bedroom. Eight of us followed. B. D. and Kahlua led the pack. When I shoved my way through the door, demonstrating leadership from the rear, I saw the flying squirrel reach the top of the curtains, take a quick assessment, and sail right over our heads, back into the hallway.

I rushed into the hall but saw no flying squirrel. We began an anxious search, looking under furniture and behind curtains. B. D. was sniffing. Kahlua was pretending to be aloof and disinterested as she stalked the unseen prey. No flying squirrel. Carol was not happy. The idea of being home alone with a dog, two cats, and a flying squirrel did not appeal. I agreed to stay around and look for the lost animal.

Still casting inquiring looks this way and that, the children prepared to leave. At least things had calmed down. Then Laura picked up her sweater. The flying squirrel glided out of a sleeve and into the dining room. More barking, more squealing, more loud advice. I finally caught the thing under the dining room table, got bitten on the hand, said some bad words, and tossed the creature into the air. In the blink of a cat's eye, it scampered up the attic stairs, back to its home.

Carol decreed we would trap no more flying squirrels. Being an ecologist who knew he had been defeated, I declared we should learn to live with our native wildlife anyway. We still listen to flying squirrels all night. The children still wonder what they will learn about them next. One captivating feature of the field of ecology is that we don't need to venture into a rain forest or tropical jungle to observe the interaction between plants, animals, and their environments. Children, especially, should be encouraged to find out more about animals or plants that interest them, especially those that live around them.

Gray squirrels are familiar to most people in the eastern United States because of their ability to thrive where people live. Scientists know more about urban species like gray squirrels than ever before, but unsolved mysteries still surround such everyday species. For example, gray squirrel populations occasionally increase to high densities in an area and then make mass migrations. This does not mean half a dozen individuals decide to move a mile or so through the forest. Records exist from the 1800s of thousands of squirrels moving overland in the same direction. The trips sometimes ended when the squirrels reached a large river, such as the Ohio or Mississippi, that turned out to be wider than a gray squirrel can swim. A few recent records exist of smaller migrations. Busy highways, rather than rivers, now prove to be effective migration barriers. Most ecologists attribute such movements to food scarcity after periods of food abundance and increased population sizes, but the precise workings of the phenomenon remain equivocal. For example, how does an individual know which direction to go? Gray squirrels are one of the most obvious wild mammals living in eastern North America. They can probably be seen cavorting in front of the biology department on every college campus in the region, and yet not all ecologists agree on why mass migrations occur.

Scientists do not know nearly as much about the basic ecology of our native plants and animals, even the most common ones, as the public may think. Our children need to know this. They need to be aware that the challenge of understanding the ecology of species that share the world with us will be a lasting one. Mysteries surround even a group of animals that any child can distinguish from all other animals: turtles. Most people have seen a turtle crossing a road at one time or another. Sometimes it is a box turtle, the high-domed yellow and black kind that can completely close its shell. Box turtles live on land, so when we see one on a highway, we can assume it is just walking to the woods or field across the street. But what about aquatic turtles that cross roads. Where are they going?

Unless it is a female turtle looking for a place on land to lay her eggs, the chances are that an aquatic turtle is traveling from one body of water to another. How does it know where another lake or stream is? In a recent discovery, ecologists determined that turtles can definitely locate water, even if it's a mile or more away. But no one knows for sure how they do it. Do they smell it? Do they hear frogs calling? Do they see it in some way, such as by light reflected from the surface of the water? One thing is certain: They do not just wander aimlessly. They can walk directly from one aquatic home to another, through woods and across fields.

Why is it important to understand how a turtle finds its way from one body of water to another? If we are to fully appreciate the effect of human environmental impacts on species, we must understand the behavior of species under natural conditions. Animals do fascinating as well as mundane things. The more behaviors we are aware of, the better we will understand the species in question. Yet here we are, after decades of biological studies, still ignorant about something as ordinary as how a turtle knows where it's going.

Another turtle mystery involves the chicken turtle, which lives in ponds and lakes in the southeastern United States. The chicken turtle has both a physical characteristic and a behavior quite different from other turtles in the same area. The neck of a chicken turtle is as long as its body, twice as long as that of other turtles of the same size. Why do chicken turtles have long necks? Do they strike out at fish that swim by? You might think the function of something as obvious as an abnormally long neck in a common native species would be understood. But no one knows for sure why chicken turtles have long necks.

The egg-laying behavior of chicken turtles is another ecological riddle. North American turtles typically lay their eggs in the late spring and summer. Sea turtles come ashore on the beaches on summer nights. Snapping turtles, painted turtles, and slider turtles normally lay their eggs from April to August. But chicken turtles lay their eggs on warm days in the late fall and winter. This is a true ecological mystery. Why does one of the many species of turtles in North America lay its eggs at exactly the opposite time of the year from when the others do? Such unexplained natural phenomena bear explanation if we are to have a thorough appreciation and understanding of the world around us.

America's children are better educated about environmental issues and the science of ecology than ever before. Maybe with this knowledge their generation will be able to manage the earth's natural habitats and their inhabitants in a proper manner. Nonetheless, school textbooks and TV nature shows have a common failing. Both leave children and adults with the impression we know a lot more than we really do about the ecology of unusual as well as common animals. Before we become smug and complacent about how environmentally sophisticated we are, remember our level of ignorance about everyday animals. Modern technological tools allow us to probe into biological mysteries that were not even hinted at fifty years ago. Using these tools, ecologists constantly discover details about complex behaviors, and ecological geneticists reveal previously unknown relationships among species. And examples of the dependency that certain species have on specific components of their environment and on other species continue to be revealed.

A fundamental question in the science of ecology is, How does a particular type of plant or animal interact with its environment? Children should be encouraged to ask why a species does things one way rather than another. This is an elementary way to look at the field of ecology, but it is what ecological research amounts to in many instances. Simply walking through the woods, wondering about how the plants and animals exist or reflecting on when they are evident and when they aren't, addresses the same kinds of mysteries that research ecologists deal with daily.

Children need to realize that they can potentially influence our knowledge of the natural world. They also need to understand that they can, and should, get involved in environmental issues. One way for children to raise their level of environmental consciousness is to read *50 Simple Things Kids Can Do to Save the Earth,* a book written by John Javna and the staff of the EarthWorks Press. Any reading-age child should be able to get some worthwhile messages from it. So should their parents and teachers.

For example, the point is made about acid rain that the sky has more in it than meets the eye and that some of those things are harmful to the earth. The book explains that coal-burning power plants and gas-powered cars release gases that can mix with rainwater to make it acidic. Several suggestions for reducing the amount of acid rain are given. One is to drive our cars less. Another is to reduce the amount of energy we use, thus lessening the amount of coal burned by power plants. Clearly most children under sixteen do not drive cars, nor do they have much control over the use of generated electricity. The point is to instill in today's youth, who will be tomorrow's consumers, the understanding that the world's environments pay a high price for each individual's overconsumption. The issue is not whether the individual can afford it, but whether the world can.

The book suggests ways to reduce the overuse of lights in the home, the most conspicuous being turning out lights that are not in use. It also recommends using natural light when possible for reading and other activities around the home. Suggestions on how to save heat energy in the home include turning the thermostat down and wearing a sweater, using hot water sparingly, and looking for tiny heat leaks around your house. Recycling, the effects of littering, and planting trees also receive due attention, with facts and recommendations a child can understand. Some of these may sound like rather obvious proposals, but I'm not sure that middle-class America teaches this lesson any more. Too often the only expense we consider is whether we can personally afford the luxury. One theme of the book is that the cost for the unnecessary use of energy is paid for by the environment, and therefore by all of us.

Overall, the book's ideas are sound. Today's children need to become conservation-minded, though some people, especially those in certain commercial fields, may not think highly of the book's content. And not everyone will agree with all the advice and recommendations given in the book. For one thing, the book does not convey the complexity of some issues. We need to teach children, not to mention many adults, that environmental issues often do not have simple solutions; they frequently require much thought and deliberation. We should emphasize that the process of making carefully considered decisions, often through compromise, will be more effective than making hasty judgments that polarize people on an issue. Putting oneself in the position of someone who will be abruptly affected by an impetuous, though well-meaning, environmentally sound decision is a useful strategy.

An in-depth attitude change is emerging among the country's youth. Most understand that action, as well as thoughtful deliberation, is part of the process of making effective change. Some colleges have adopted an idea that is admirable because of its effect on and reflection of environmental attitudes in the educated youth of the country. The first document some college students sign upon graduation reads as follows: "I [your name] pledge to investigate thoroughly and take into account the social and environmental consequences of any job opportunity I consider." Had you graduated from Humboldt State University in Arcata, California, in any year since 1986, you would have had a decision to make, along with the rest of the seniors. Would you sign the pledge or not?

The pledge, which is voluntary, is provided at commencement ceremonies as a document that students can take or leave. Some who pick up the pledge document sign before they flip their tassels. Others take it home and sign it, and others take it home and don't. Nothing is expected or demanded. But an attitude is present, and it is one we need in our young people. More than thirty other universities, including Stanford and the Massachusetts Institute of Technology, have followed Humboldt State's lead. The students at Agnes Scott College in Atlanta have now presented their version of the pledge option, and at least two high schools in California have adopted the idea.

The pledge idea at Humboldt State was initiated by students and community members. With the support of the faculty, they managed to make it tangible. The concept makes a real statement about young people's attitudes toward how we manage our environment. As the Humboldt State students put it, the pledge is not meant to be "a once-a-year event at graduation, but a tool for exploring personal impact on our social and physical environment."

This kind of thinking probably unnerves some people. Should college students be concerning themselves with the environmental effects of a business run by experienced adults? Well, yes, I hope so. After all, these will be the men and women running the businesses in a few years. But, the concerned business leader might ask, what about the next step? Will students soon be required, rather than simply permitted, to sign such a pledge at graduation? Probably not, but that would lead to an interesting situation. Only those graduates underhanded enough to break a pledge would be the ones who would take jobs with certain companies. And is there anything wrong with asking someone "to investigate . . . the social and environmental consequences" of a business before contributing to its success?

The pledge idea calls attention to the community spirit needed today on environmental issues. Nevertheless, I imagine the reaction of some people is, so what? What difference will it make? Except, of course, that those foolish enough to sign such a pledge will not take some of the good jobs, and those who do not sign will have more options. The paradox is that those who sign such a pledge and mean it would not be turning down what they consider a "good" job. And companies that are changing their environmental practices and want to improve their public image might prefer to hire those who have signed, thus giving an advantage to the signers. I can envision a future in which companies would seek out graduates who have signed such a pledge.

Of the countless things today's children, as well as adults, can accomplish in the interest of environmental prosperity, none surpasses that of developing a staunch appreciation of the natural world. Rare and unusual creatures are always fascinating, but every animal and every plant has its unique ecology and can be an absorbing subject for ecological investigation. Ecology is a far more complex science than physics or chemistry, because it incorporates not only physical and chemical phenomena but biological ones as well. The field of ecology remains filled with intrigue and unpredictability. Ecological mysteries are waiting to be solved, and anyone might solve them.

We should teach children to enjoy nature for its own sake, encourage them to appreciate the explanation of any natural trait of plants or animals, to observe the natural world around them, to question and marvel and ponder. All of us can lead richer lives from becoming more aware of our environment, but society will profit on a long-term basis if children become knowledgeable about ecology. It is the children who will inherit the earth. We have a responsibility to teach them to cherish their natural heritage.

It's Time for
an Environmental
Attitude Adjustment

One reason that wild populations of native species on every continent are disappearing faster than they can replace themselves is that commercialization has outstripped environmental protection. Timbering of old forests is allowed to proceed at a rate faster than clear-cut land can be replaced by equally old forests. Water projects are undertaken with no controls or consideration of the long-term shock to the natural environment. The impact of such environmental abuse can be far more detrimental to the welfare of humans than the absence of an artificial reservoir. Industrial pollution destroys the integrity of ecosystems throughout the world. Wild animals are collected commercially for food or the pet trade without proper controls. If we do not take a responsible approach soon, future generations will justifiably view us as thoughtless environmental miscreants.

The situation with the common snapping turtle offers a prime example of a problem and what can be done to solve it. Snapping turtles, which contribute to the ecological functioning of aquatic habitats as scavengers, are harvested by the thousands in North America for sale in restaurants. With the exception of eating an occasional fish or duck, and biting the fingers of people who pick them up the wrong way, snappers rarely harm human interests. Yet they are being systematically and ruthlessly destroyed, and few federal or state regulations ensure the continued existence of these animals.

Now you do not have to like snapping turtles or even be environment-minded for this to be an issue of concern. All you need to be is a taxpayer. People are being allowed to harvest your natural resources for their own profit without obtaining a license or paying any fees. Snapping turtles are only one example. The same is true for many native wildlife species. Very few states have identified and responded to this problem. Laws to protect wild animals are difficult to pass because of special interests, usually related to someone's making money. This must not continue. We must develop better laws to protect all our wild animals and plants, not just those in which humans have taken a special interest. Here's how.

We should prohibit the wholesale removal of any species of animal or plant from the wild unless it can be shown that regional populations can sustain the removal rate and replace themselves. Laws like this already protect game species such as white-tailed deer, rainbow trout, and ruddy ducks. When a game species becomes scarce, regional hunting and fishing are reduced by state and federal game officials. Also, those who hunt and fish pay for the right to do so. The laws protecting game species are generally effective.

We now need to protect all wildlife, nongame as well as game species, plants as well as animals, regardless of whether they are perceived as rare or endangered. At present, in order to halt harvesting in most states, proof is first required that a species, such as the snapping turtle, is threatened or endangered. This is the wrong sequence of events. We should require evidence that a species is not—and will not be—threatened *before* any harvesting can begin. In other words, the evidence must show that populations of a species like the snapping turtle must be able to not only sustain themselves in the short-term but still be doing fine half a century later. The burden of proof should be on the harvesters, not on those who want to protect the species. Also, people who benefit financially from these natural resources should be charged for the privilege.

One lame argument that has been used to lobby against more stringent controls on removal of native species is that children will be discouraged from enjoying nature if they cannot catch their own lizards, frogs, or caterpillars. This can be easily remedied with a law that allows anyone to keep up to four specimens of most animals but prohibits the sale of native wildlife. Why should anyone be allowed to remove our native wildlife and sell it without compensating the rest of us? If children are really to enjoy nature, we must ensure that nature stays intact for them.

Each year, millions of dollars change hands in the United States and other countries as our ever-diminishing natural resources are sold for food, pets, or other uses. The plants and animals of the country belong to everyone. Let's adjust our environmental attitude by recognizing that our natural wildlife is as valuable as any resource we have. It is an adjustment that is several decades overdue.

Land development, timber operations, water resource projects, and industrial pollution should all operate under the assumption that the status quo is preferable to change. Today, if someone cannot justify why an environmentally destructive activity should be stopped, the project can proceed. Let's turn the formula around: Prove that a project will not result in a net loss of native biodiversity or natural habitat before approval is received. We have reached a point in our environmental decline when we must forfeit laissez-faire commercial progress if it conflicts with environmental security, which is indisputably more critical to our long-term welfare.

Each of us individually can help preserve biodiversity and protect the natural habitats that still surround us. Set a good example (to children as well as adults) by recycling, by moderating consumption, and by rejecting products acquired through environmental destruction. Make local, state, and federal officials aware that natural habitats and all species of plants and animals are not limitless resources and that our rash exploitation of them cannot be continued. Support the efforts of research ecologists who, in the course of unraveling the infinite variety of ecological interactions found on this planet, also reveal direct and indirect benefits of other species to our own. All of these individual endeavors involve a common thread: to be environmentally responsible and educate others to be likewise.

We must also control human population growth. Perhaps those who reject certain means of birth control will consider the subject from another perspective. Instead of condemning birth control measures, they should justify their support of policies that are destroying the rest of the world's species and natural habitats. Adding more people to an overpopulated planet will result in an inexorable decline in the numbers of individuals in wildlife populations, the number of populations themselves, and eventually the number of species. Although we can refuse to address the issue of birth control, we cannot avoid the consequences of that refusal. Other species will not be the only ones to suffer if we fail to limit human population growth.

Epilogue: The Way the World Could Be

I began this book with a rather grim view of how the world could be a hundred years from now. I end it with another vision—one I think most of us would prefer.

Fewer people will live on the planet Earth than now, but they will have a far more secure existence and a higher quality of life. Society, including governments and religious organizations, will have realized that humans can limit their own population in much kinder ways than the raw controls imposed by nature. A global standard will place boundaries on population expansion.

The success of the natural world, and of human beings, will rest on an environmental attitude that has resulted in laws and actions in the best interest of all species, including our own. The sun's energy will be better harnessed, and people will no longer rely on nonrenewable energy sources. An attitude that excessive consumption is both inappropriate and unnecessary will pervade society. The distribution of resources will be equitable on a worldwide scale.

Above ground, the buildings will be taller than buildings of today. Cities will also extend underground. Humans will have discovered the obvious, that cities do not need to take up vast amounts of land surface area. Many of today's sprawling metropolises will have contracted. A city like Phoenix will be built up and down instead of out, sparing the natural desert habitat, where such native life-forms as saguaro cactus and diamondback rattlesnakes will provide welcome proof that life has conquered, and depends on, even the harshest (by human standards) environments. All natural habitats will be accorded economic value commensurate with their true worth, and none will be considered cheap or indispensable.

In this new time, rainfall will be pure. The sun will shine through a pristine atmosphere, to be enjoyed around the globe beneath a healthy ozone layer. Children will once again laugh and play in crystal clear streams unspoiled by erosion runoff or industrial pollution. The abundance of native wildlife will proclaim the absence of unseen pesticides, herbicides, and other chemical contaminants.

The world's leaders will have agreed that certain regions of the earth, including all oceans and both tropical and temperate forests, are inviolable natural areas to which no one has special commercial rights. International environmental treaties will prohibit exploiting any species beyond its ability to recover naturally and will acknowledge that natural habitats and their organisms are the property of all the earth's inhabitants.

The land and waters will be recognized as sacred trusts, to be enjoyed in the present yet preserved for future generations. Waste-free oceans of blue and green will wash on the world's beaches. Anyone will be at liberty to swim or sail without concern for human contaminants. Dolphins and whales, free of unnatural predation by the human species, will be anticipated sights. The plants and animals will be protected, free from the caprice of, or abuse or destruction by, the person who holds title over the land.

Elected officials will no longer worry whether supporting environmental projects would be a smart political move. They will instead promote "ecosystem values." The idea of destroying the environment or bringing any species to extinction will be totally unacceptable. All plant and animal species will be protected from unsustainable human removal. Elephants and tigers will again be keystone species in their jungle homes. Alligators and crocodiles will bask along riverbanks in the security of their recognized right to live their natural lives. No one would even consider lobbying for programs that jeopardize the existence of a species in its natural habitat.

Nature preserves—big and small—will be ubiquitous and will include all habitats native to a region. Wildlife in nature sanctuaries will live in totally natural environments and enjoy non-negotiable protection from human activities. Nocturnal animals will experience nights in which darkness depends only on the phase of the moon and cover of clouds; no artificial lighting will disturb the shadows. Sea turtles will dig their nests in the beauty of beaches lit only by the moon and stars. Their hatchlings will enter a night surf shining with the greenish glow of phosphorescent ocean life. No invasive artificial noises of humans will be heard in the sanctuaries, only natural sounds: the howl of wolves in the northern forest preserves, the chattering of monkeys in protected tropical rain forests, the songs of birds throughout the world.

Easy access to the natural habitats will be possible for those who choose to experience them. Those who visit a tropical rain forest will walk beneath enormous trees that tower overhead, beneath a canopy so dense that daylight becomes darkness on the forest floor. Iridescent butterflies will flit along the path, amid a captivating array of flowers while colorful birds fly above. The sound of chain saws and bulldozers and the sight of fires and smoke will be forgotten images of an earlier century.

Some people will still prefer to remain in the cities, but all will have been taught from early childhood that natural habitats and ecosystems are vital to the survival of humans as a species. Those who choose to live away from the cities will do so with the understanding that they are part of the ecosystem, not the rulers of it. Everyone will understand that humans are encouraged to enjoy nature but not permitted to disrupt it.

How will the human traits of greed, self-centeredness, and aggression affect the environment? How will society handle the fact that some people are more industrious, more innovative, more ingenious than others? These attributes will always be with us as a species, but industry, innovation, and ingenuity will be redirected toward our learning to live with the earth instead of acting in ways that pit us against it. Just as all societies teach intolerance of theft and capricious murder, lifelong environmental education programs will teach that abuses against the natural world are equally unacceptable. Unjustified or unwarranted human destruction of natural habitats or native species will be viewed as a crime against nature and will not be tolerated.

People will still have houseplants, gardens, and pets. But guidelines dictating the sale and distribution of organisms will result in no net loss of native wildlife. The rules will be accepted throughout the world. Not to accept them will be considered unethical and immoral. People will have bird and animal feeders in their yards and will be pleased that myriad creatures are around to enjoy them.

Ecological research will continue in depth on species in their natural habitats. Each revelation will contribute not only to an understanding of nature but also to an understanding of ourselves. Studies of natural history will have demonstrated that scientists studying the lives of plants and animals are dealing with the mysteries of an entire universe. No one will be dismayed that life on earth presents more ecological puzzles than can ever be solved. Everyone will cherish the knowledge that a multitude of plants and animals are still on earth, and that we are around to study them and to enjoy them as they are.

Isn't this is a better vision than a world without wildlife?

Index